MW00791162

SATURN I/IB

THE COMPLETE MANUFACTURING AND TEST RECORDS

Alan Lawrie

An Apogee Books Publication

Dedication

To Olwyn Georgina.

Tramps like us, baby we were born to run.

All rights reserved under article two of the Berne Copyright Convention (1971).
We acknowledge the financial support of the Government of Canada through the Book Publishing Industry Development Program for our publishing activities.

Published by Apogee Books, Box 62034, Burlington, Ontario, Canada, L7R 4K2.
http://www.apogeebooks.com

Printed and bound in Canada
Saturn I/IB - The Complete Manufacturing & Test Records
by Alan Lawrie

DVD material provided by MSFC, designed by Robert Godwin

ISBN 9781-894959-85-8

©2008 Alan Lawrie

All numbered photos are from NASA (either via NARA or from the NASA web), whilst un-numbered photos have been supplied by Alan Lawrie

Foreword

The author of this book, Alan Lawrie, herewith completed his documentation of the Saturn Program. He has previously published a book on the Saturn V, which presents impressive detail on that program.

He has now undertaken to publish a similar book on the Saturn I and Saturn IB, which were the precursors of the Saturn V. The Saturn Program was already started under the U.S. Army. The Commanding General John Bruce Medaris saw the need for much larger launch vehicles than were available at that time, or even in planning stages. Under the Horizon Project he wanted to establish a Lunar Base from where to observe what happened down here on Earth. Spy satellites had not been invented at that time. This idea created the need for much larger launch vehicles.

For this reason a Future Projects Group in the Army Ballistic Missile Agency had already undertaken a number of studies which progressed step-by-step to a lunar launch vehicle. When the majority of the Rocket Team transferred from the Army to the newly established NASA, I decided to leave the Army and to join the Marshall Space Flight Center, especially since I was offered the job of Deputy Manager of the Saturn Program under Dr. Oswald Lange as Manager. A few people stayed with the Army, among them Arthur Rudolph. In my new position I spent a lot of time at the Missile Division of the Chrysler Corporation at the Michoud Assembly Plant, near New Orleans. I also made frequent trips to the Kennedy Space Center to assure that the new launch facilities could meet all the requirements of the new Saturn launch vehicles.

The Saturn I was born under an initial Advance Research Projects Agency (ARPA) contract which called only for static firings at Marshall's test facility to study the performance, feasibility and reliability of a large engine cluster which would permit a big jump to a much higher thrust level of a launch vehicle stage. Since all static firings showed excellent performance and reliable operation of the eight engine cluster, it was decided upon President Kennedy's announcement to go to the Moon, to use this configuration also in an actual launch vehicle stage, to be named S-1 stage of the Saturn I.

The Saturn I as well as the Saturn IB were used in early missions to prepare for the final operation of the Saturn V which did not become available for several years. This new requirement for a lunar mission made it for time reasons also mandatory to skip all of several intermediate steps and to take a huge jump directly from the Saturn I to the Saturn V.

After Arthur Rudolph had completed the development and production of the Pershing System, he left the

Konrad Dannenberg with Alan Lawrie

Army and rejoined the Rocket Team in August 1963. Based on his extensive experience in the production of rockets in the Mitttelwerk in Germany and in this country with the Pershing, Wernher von Braun appointed him as the Manager of the Saturn Program. Arthur Rudolph selected as his deputy an Air Force Colonel, and I worked, from then, on a Saturn-based Space Station, and other Post-Apollo programs.

The problem with the Saturn I was the lack of a second stage. After many considerations and calculations it was decided to build a new hydrogen-oxygen propelled second stage using a six engine cluster of the RL10. Since this stage was not yet available, von Braun decided to make the initial launches just with the first stage and a simulated dummy second stage. This step-wise approach of firing initially only the first stage was also the most desirable development progress in the eyes of the rocket team, as well as for Wernher von Braun. These one stage tests demonstrated the excellent performance of this new powerful first stage engine cluster. As soon as the S-IV stage became available, it was used in additional launches of two-stage Saturn I's.

The thrust level of this new second stage was rather low at 90,000 pounds of thrust. It was therefore decided to use the much more powerful J-2 rocket engine of 200,000 pounds of thrust as soon as it became available. It became known as the S-IVB stage. With the more powerful thrust level, the ratio between the thrust levels in the first and second stages was much more advantageous, and the Saturn-IB became a very efficient launch vehicle which conducted a number of early Apollo Missions. It could not transport the lunar payload to the moon, but all the lunar landing equipment could be put into an earth orbit, and many procedures of the landing preparations could be demonstrated with this combination. The Saturn IB had also the purpose to test the new J-2 engine in actual space flight.

In hindsight it can be said that the entire Saturn Program was apparently the most efficient way to go to the

Moon for a landing and a safe return of the crew back to Earth. In conclusion, the author describes all these many steps of the Saturn development in much greater detail than it was done in this summary foreword.

I am happy that this Apollo-Saturn technology is being used again for the Ares 1 and V programs to return to the Moon and to eventually proceed to Mars and other places in the solar system.

Konrad Dannenberg, Former Deputy Head of the Saturn Program at NASA Marshall Space Flight Center.

Acknowledgements

The Saturn collection at The Salmon Library at The University of Alabama at Huntsville continues to be a fine, well organized collection of Saturn-related documents primarily established as the source material in the research for the NASA history, Stages to Saturn. Anne Coleman and her colleagues Gary Glover, Sharmila Rajasekaran and Vignesh Subbian again provided helpful, friendly assistance keeping me well stocked with photocopies and refreshments.

Anne Coleman

Arlene Royer with von Braun's Weekly Notes

The UAH team

Since my research for the Saturn V book the National Archives, southeast region, have moved to a new facility at Morrow, Georgia. The new building is clean, modern and architecturally pleasing to the eye. Arlene Royer has taken over responsibility for the collections of documents shipped across from the Marshall Space Flight Center. The documents arrive in cardboard boxes and when time allows the documents are restored, sorted and indexed. The situation has improved significantly since my first visit four years ago and Arlene is to be congratulated on her fine work. The wealth of original Saturn documentation at NARA is staggering and on my visits I was able to unearth significant Saturn documents that are not available anywhere else. Arlene also has the original von Braun weekly notes and the original Saturn still photo negatives, stored in the refrigerator. Arlene kindly provided me with a number of photographs from the refrigerator for inclusion in the book. Thanks also to Arlene's assistant Shane Bell who located some useful documents for me.

Don Brincka was in charge of the testing at SACTO throughout the Saturn era. Don has a wealth of firsthand knowledge of the Douglas activities at that site

Don Brincka

and was invaluable in assisting me in the preparation of this book. Don took me on a tour of the old test stands at SACTO and we sat in his old office that he probably had not gone into in 30 years. Over the course of the past two years Don has painstakingly and meticulously answered the many obscure questions I have flung at him. In addition he provided me with many original Douglas documents and reports. Thank you so much for your help. Look out for Don on Discovery's Moon Machines series.

Vince and Gail Wheelock

Vince Wheelock is what can only be called the Rocketdyne historian, although he is now retired from the company. Vince has an incredible library of rocket engine and space documentation, photos and books as well as maintaining the Rocketdyne Leadership and Learning Center on a voluntary basis. I first got to know Vince after the Saturn V book came out. Since then we have communicated regularly during which time Vince has spent a lot of time answering my many questions and providing me with many original documents. Many of the photographs in this book came from Vince. Olwyn and I were lucky enough to visit Vince and his wife Gail in January 2008 at their home in Los Angeles. Gail is the organizational mastermind behind the library. Vince is another without whom this book would have many gaps. Thank you both for your help and friendship. Check out Vince's AIAA book on Rocketdyne.

Bob Jaques

Bob Jaques is a former historian at MSFC and has helped me greatly over the years. Bob was kind enough to show me his wonderful personal collection of space memorabilia.

Gene Robinson worked at SACTO and unearthed some unique photographs from his collection which he kindly allowed me to use in this book.

Ralph Allen is the Historic Preservation officer at Marshall and as such knows every nook and cranny of the

Ralph Allen

facility as well as the history of all the test stands. Ralph provided me with many photographs, answers to questions and took me on a tour of Marshall which included hacking our way through the forest to the old Power Plant observation bunkers and a rummage around the MSFC bone yard looking for old pieces of Saturn rockets and engines. Ralph and his wife Linda very kindly gave up their Sunday to take me on an educational tour of Huntsville where we looked out the former houses of the von Braun team members.

Dave Hewitt

Ray Byrd worked in the VAB at KSC and has some wonderful stories to tell about what really went on there.

Dave Hewitt at MSFC helped locate some obscure test reports. Dave shows an enviable enthusiasm working on the next generation of moon vehicles. Watch out for him on NASA TV.

David and Mrs Akens with Alan Lawrie

David Akens was the Chief Historian at MSFC throughout the 1960s. He wrote the immensely important Chronologies of MSFC and the Saturn Illustrated Chronologies. David and his lovely wife kindly invited me to their house in Huntsville during my research.

Duncan Copp and Chris Riley made the most enjoyable films on the Apollo program – the film In the Shadow of the Moon and the TV series Moon Machines. Their attention to detail is to be congratulated and the Saturn sections are of particular interest. Check these out if you have not done so already. You are also recommended to purchase the entire catalogue of Mark Gray's Spacecraft Films DVDs.

Yvan Voirol with Konrad Dannenberg

Konrad Dannenberg was one of von Braun's original team and was deputy head of the Saturn program. Konrad kindly invited me to talk at the 50th anniversary of the Explorer satellite launch held in Huntsville in January 2008. Konrad has so much experience and enthusiasm it is quite staggering. Thank you so much for being kind enough to pen the Foreword to this book. I cannot think of a more appropriate person. Thanks also to his wife Jackie for making the Explorer celebrations such a success.

Mike Wright is the MSFC Historian. Mike was very kind in providing me with various obscure CDs and DVDs and for supporting my tours around Marshall.

Mike Jetzer is someone who has gone out of his way to support the research in this book. I met Mike at the 50th Anniversary of the Explorer launch in Huntsville and he has been a constant source of encouragement.

Yvan Voirol accompanied me on one of my visits to Huntsville and we had an enjoyable time visiting space sites and meeting up with former members of von Braun's team. Thank you for your valued friendship. Check out Yvan's space memorabilia web site.

Pauline Roe and Karen Bartha and Bernardo Martins at EADS Astrium kindly provided research assistance.

Jesco von Puttkamer at NASA HQ allowed me to use personal photos he took in Huntsville in 1969.

Judy Claussen and Mark Knight at Premier Turbines provided an informative insight into the largely-forgotten Neosho rocket facility and kindly allowed me to use some of their photos that originally appeared in their book, The first fifty years in Spacetown USA.

Thanks to the British Interplanetary Society for their Spaceflight magazine and for allowing me to submit the occasional Saturn article.

Thanks to Steve Smyth who took some photos in the VAB over 30 years ago and allowed me to use one of these. Steve was kind enough to take the trouble to locate the original negative from 1973.

Thanks to Robert Pearlman for his CollectSpace web site which I probably spend far too long reading each day. This is a tremendous resource and Robert is a true gentleman.

Bob Foust who worked for many years on the RL10 rocket engine at Pratt & Whitney and now works at The Aerospace Corporation provided valuable help on the RL10.

Paul and Barbara Coffman

Paul Coffman worked at Rocketdyne and has helped me since the earliest days of my research. Thank you for your continued help. It was a pleasure meeting up again with both you and your wife Barbara. Thank you.

David Christensen

Dave Christensen is an ever present whenever I visit Huntsville. Dave has Saturn experience back to the earliest days and was instrumental in setting up the Saturn collection at the University of Alabama at Huntsville. Thank you for your continued friendship and support.

Thanks to Kipp Teague for allowing me to use some photographs from his on-line Project Apollo Archive. Also thanks to Phil Broad for use of photographs from his on-line site.

Thanks to Frank Seabourne, Chairman of Automotive Importing Mfg., Inc. who allowed me to visit his warehouse which happens to be the former Vehicle Checkout Laboratory at SACTO. Thanks also to Aerojet for allowing me to visit the rest of the SACTO site.

Tom Hancock works at Marshall and does a lot to promote space in the community. Thank you Tom for running the Anniversary celebrations in Huntsville in January 2008 and for supporting my Saturn activities.

Thanks to Rob Godwin for publishing this book and the others in the Apogee series.

Thanks to Olwyn Georgina and my parents without whom this would not have been possible.

Introduction

When I did the research for the companion book to this one (Saturn V – The Complete Manufacturing and Test Records, Apogee, 2005) I made sure that I did not lose sight of the earlier Saturn rockets that paved the way for the Saturn V. Consequently by 2005 I already had a wealth of information on the Saturn I and Saturn IB rockets. It was clear that without the tremendous groundwork performed by NASA and the US industry on these earlier programs the Saturn V program would never have been as successful as it was and men may not have walked on the moon by the end of the 1960s, as required by President John F Kennedy in his famous speech in May 1961.

After the Saturn V book I established many new contacts as I continued my research activities, homing in specifically on the earlier rocket series. The intention was to publish a companion volume to the Saturn V so that the complete history of all the Saturn rocket stages and facilities was preserved. Finally, after a further 3 years, this book has seen the light of day.

In the intervening period I have visited new locations as well as revisiting many of the places I went to for the first book. This time around I was lucky enough to tour the remains of the Sacramento test site where the S-IV and S-IVB stages were test fired, in the company of the Saturn test director at the site in the 1960s. It is incredible that the site has remained almost untouched for over 30 years. I went back to the Marshall Space Flight Center, where the Saturn program was run from, as well as re-visiting the archives at the University of Alabama in Huntsville. Although I had previously been to the official repository of Saturn documents, The National Archives in Atlanta, when I visited their new location and saw how access and indexing had improved so much I was so overwhelmed that I made a point of returning again within a few weeks.

Many people have gone out of their way to help me in the research for this book and to make the information contained within as complete and accurate as possible. I take my hat off to all of these fine people who are mentioned individually in the Acknowledgements section. I apologize if I forgot anyone.

As with the Saturn V book I decided to cover the life history of each Saturn stage. The only difference was that this time I decided to continue the story up until the launch, whereas in the earlier book I stopped at the delivery of each stage to KSC. The main reason for this was that I thought that the processing at KSC was of interest, especially as there were stages that travelled back and forth to the Cape. I decided to allocate more space to the sections covering facilities, transportation and engines as these deserve adequate recording as they are very poorly covered in the existing texts. Also I gave greater weighting to some subjects that I consider have been particularly under reported in the past, such as SACTO.

Inevitably, as the activities described were performed even earlier than those on the Saturn V, and now almost 50 years ago, memories of individuals and the locating of documentation was harder than with the previous book. However, I am confident that I have been able to present an accurate picture of events. This was largely due to the enormously generous provision of early and original Saturn documents by individuals and the location of others in the archives. I attempted, where possible, to go back to original source documents for the research. Even then there were occasions when official documents contradicted one another in details such as firing durations or dates of shipment etc. In these cases I have tried to establish what the correct version was. Any errors or mistakes that remain are my responsibility.

I believe that, with this book, I have managed to bring together all the key Saturn I and IB records for the first time. As time marches on this opportunity would have become harder than it already was. I am particularly pleased that some of the more obscure activities will be better recognized for their important contributions. The RL10 activities at Pratt & Whitney, the Rocketdyne activities at Neosho, the S-IV testing by Douglas, the barge and Guppy transport activities and the many H-1 engine issues fall into this category.

I hope that you find this book to be a useful reference of the pioneering rocket activities of nearly half a century ago and that, together with the earlier Saturn V book, the story of the manufacturing and testing of these magnificent machines is now recorded for posterity.

Alan Lawrie

Hitchin, England
July 2008

Contents

NASA, George C Marshall Space Flight Center (MSFC) - Huntsville

The George C Marshall Space Flight Center (MSFC) was activated on 1 July 1960 with the transfer of personnel and facilities from the US Army Ballistic Missile Agency. The Saturn I/IB programs as a whole were managed from NASA's MSFC in Huntsville, Alabama under the direction of Dr Wernher von Braun.

The East Test Area included the two-position Saturn I/IB Static Test Tower (facility 4572). Originally the stand comprised of a West side only. The Test Stand was built in 1957 and used to conduct 487 tests involving the Army's Jupiter missile. On 14 January 1959 ABMA began modifying the East side of the existing Static Test Tower for the static testing of Saturn I stages. On 4 February 1963 MSFC started conversion of the West side of the Static Test Tower for F-1 engine firing capability. In November 1964 the West side was converted for the additional use of S-IB stages, a task that was completed in February 1965. However this side of the tower continued to be used exclusively for testing of F-1 engines. Conversion of the East side of the Tower for S-IB stage testing was completed in February 1965. The original East side was used for all test firings of Saturn I/IB stages. The West side was used in the 1980s to support Space Shuttle solid rocket motor tests. The test stand is no longer used but is still in place with the SA-T stage lying next to it. The 175-foot facility was selected as a National Historic Landmark on 3 October 1985.

The Power Plant Test Stand for testing single H-1 engines in a vertical orientation was located next to the Static Test Tower. It was originally built by the Army in 1954 as a horizontal test stand. In 1956 it was converted to a 110 feet vertical power plant test stand for the H-1 with two static firing positions. There was a separate blockhouse and the stand was capable of firing an H-1 engine for 160 seconds. The test structure (facility 4564) was demolished in the 1970s and replaced by the Transient Pressure Test Article Test Stand, constructed in 1987. This test stand bears the same building number and was used to verify the sealing capacity of the redesigned Space Shuttle solid rocket motors following the Challenger explosion.

However, two observation bunkers for the original Power Plant Test Stand remain. The original bunker is a small metal chamber with a hemispherical dome built around 1954. The replacement bunker (facility 4560) was designed by the Miami Architectural-Engineering firm of Maurice H Connell & Associates and was constructed in 1958 on top of a mound for better visibility.

Three views of the Power Plant Test Stand (Late 50s)

Nearby, in the East Test Area, the J-2/battleship test stand (facility 4514) was constructed. This test stand has been demolished.

Construction of the S-I/IB Dynamic Test Tower (building 4557) was completed on 17 April 1961 and the facility was handed over to MSFC Test Division. Modification of the tower for Saturn IB testing was achieved in January 1965. This tower was demolished on 15 July 1996.

Original Power Plant Test Stand observation bunker (2007)

Later Power Plant Test Stand observation bunker (2007)

J-2/Battleship Test Stand (1960s)

Demolition of the S-1 Dynamic Test Stand (15.7.96)

included the first stage static test stand, an F-1 engine test stand, the Saturn V launch vehicle dynamic test stand and ground support and component test positions.

The new area containing the Saturn V booster test stand and the F-1 test stand was designated the West Test Area and supplemented the existing East Test Area.

Rocketdyne – Canoga Park

Rocketdyne designed the H-1 and J-2 engines, used on the S-I/IB and S-IVB-200 stages respectively. During the active phase of the Saturn I/IB program Rocketdyne was a division of North American Aviation. Their main engineering and manufacturing facility, then as now, is located at Canoga Park, to the northwest of Los Angeles. Rocketdyne opened the 51-acre facility in 1955 and have designed and manufactured many of the nation's rocket engines. Manufacturing, assembly and rework activities on the H-1 transitioned to the Rocketdyne facility at Neosho, Missouri.

Initially, manufacturing of components and final assembly of both Saturn engines was carried out in eight buildings in the Canoga Park complex. As well as machining facilities there were two large brazing furnaces for the brazing of the thrust chamber tubes and injectors. Special areas for precision cleaning and assembly and for various types of non-destructive inspection existed. An Engineering Development Laboratory provided test facilities to support manufacturing production.

Test firing of the J-2 engines was performed at Rocketdyne's Santa Susana Field Laboratory, not far from Canoga Park, whilst the H-1 was fired at both SSFL and at Neosho.Rocketdyne was acquired by the Boeing Company in 1996 and then sold to Pratt and Whitney in 2005. Pratt and Whitney Rocketdyne has other local facilities at West Hills and DeSoto in Los Angeles.

MSFC utilized an altitude facility for firing the RL10 LOX/LH2 engine from the second quarter of 1962 onwards. This stand fired the RL10 engine horizontally with a steam ejector evacuating the exhaust until the engine started. At that time the diffuser maintained vacuum. The diffuser used a half-shell that sealed around the band on the thrust chamber that was also used for supporting the engine during certain handling operations. This was the same arrangement as was used on the earliest Pratt & Whitney test stands E1 to E4 at West Palm Beach. New Saturn V facilities built at MSFC

Rocketdyne - Neosho

In 1941 Camp Crowder (named after the old Fort Crowder) was built at the southern side of Neosho, Missouri, as a training center for the US Army Signal Corps. Up to 47,000 troops were soon stationed at the Camp, one of the Army's largest installations in the mid-west. The Camp was used to house German and Italian prisoners of war between 1942 and 1946.

The 900 building, former Camp Crowder laundry (Late 1950s)

The year after the war finished the Camp was closed as a basic training camp. The many facilities that had been established for the troops were taken over by local industry and businesses. Fort Crowder was completely deactivated in 1958 and was declared surplus property in 1962.

The Rocketdyne Manufacturing Facility (Late 1950s)

In 1956 the Army transferred part of Camp Crowder to the Air Force for the construction of a rocket engine manufacturing plant. The installation was known as Air Force Plant No. 65. Groundbreaking for the new facility took place on 21 April 1956. The plant was to be operated by Aerojet General Corporation on behalf of the Air Force, but the contractor was changed to Rocketdyne on 14 June the same year. During the Rocketdyne occupancy the facility was informally known as Spacetown USA.

Rocketdyne took over 900 Building on the south western side of Camp Crowder. It had been the laundry for the Camp until the Korean War. The 900 Building was

Dr von Braun visit to Neosho
(with W Baisley and E Wright) (17.5.61)

used by Rocketdyne between 1956 and 1968 to dismantle and store rocket engines. In parallel construction started on the test facilities and on a new rocket engine manufacturing and assembly building. The Manufacturing Facility was built in 1956 and used by Rocketdyne from 1956 to 1968 to make rocket engine parts and assemble rocket engines. Manufacturing processes performed included machining, welding, heat treatment, plating, degreasing and final assembly.

The Neosho Engine Test Area (ETA) consisted of two large engine test stands, each with two positions. Test Stand # I (North Engine Test Stand) was for H-1 production testing whilst Test Stand # II (South Engine Test Stand) was for other large engines. Each stand was equipped with a blast deflector and a concrete waste liquid drainage trough leading to a storage pond. One control center controlled both stands. The test stands were built in 1956 and used until 1968 by Rocketdyne to test fire, Thor, Atlas and Saturn H-1 engines. The first Thor engine completely fabricated at the plant was delivered to the Air Force in February 1958.

The Component Test Area (CTA) east of the Engine Test Area was where tests were performed on rocket

Engine assembly in the Manufacturing Facility.
Engine H-7054 was to fly on AS-202 (1965)

Engine assembly in the Manufacturing Facility (1960s)

Aerial view of CTA (foreground) and Engine Test Stands (background) (1960s)

components such as gas generators, turbo pumps and vernier engines. Test Cell # III was used to hot fire test H-1 turbo-pumps and gas generators prior to installation on the engines.

The first H-1 engine test at the former Air Force Plant No 65 at Neosho, Missouri, was on 25 May 1962. The first H-1 production engine to be delivered from the Rocketdyne plant at Neosho occurred in September 1962.

Construction of one of the two Engine Test Stands (1956)

One of the two Engine Test Stands (1960s)

In January 1968 NASA and Rocketdyne decided to move the site of H-1 engine production from Neosho back to Canoga Park, California. The plant at Neosho was sold to Continental Aviation and Engineering (CAE) in 1968 and the production and testing of rocket engines made way for the overhaul and repair of jet engines. In 1992 Sabreliner Corporation acquired the facility and in 2003 Premier Turbines, owned by Dallas Airmotive took over. They are the current owners of the facility, which continues to be used for the servicing of jet engines.

Currently Premier Turbines is located in the original Rocketdyne Manufacturing Building. The 900 Building, the old laundry, is no longer owned by Premier Turbines. It was used from 1968 to 1992 for the clean-

ing of engine fuel control components and now a large portion has been demolished and the remaining part is used for commercial mini storage. The Engine Test Area is part of Camp Crowder and is now owned by the National Guard. The concrete structures are still in place but all the equipment was removed many years ago. A Hazardous Waste Pit, used for burning off propellants midway between the two Engine Test Stands is now the subject of an environmental contamination study after having been drained and filled in around 1981. The Components Test Area continued to be used by the aircraft engine business until 1973 but is now

Air Force Secretary E Zuckert and delegation prepare to witness a firing (27.6.65)

An H-1 engine being fired in one of the two Test Stands
(1960s) 6518297

An H-1 engine being fired in one of the two Test Stands
(1960s)

outside the Camp Crowder boundary and is inactive. The former CTA site was sold to the Water and Wastewater Technical School Inc., in 1976.

Rocketdyne - Santa Susana Field Laboratory (SSFL)

The Santa Susana Field Laboratory (SSFL) was a rocket testing facility outside of Los Angeles that was used extensively during the development and production phases of the Saturn I, Saturn IB and Saturn V rockets. It was used for development, qualification and some acceptance testing of H-1 engines and for qualification and acceptance testing of all J-2 engines.

The site is located in the Simi Hills area of Ventura

County, 30 miles northwest of Los Angeles. The site is situated on a ridge, between 1,640 and 2,250 feet above sea level, overlooking Simi Valley to the north and the San Fernando Valley to the southeast. Close to the Rocketdyne manufacturing facility at Canoga Park, SSFL has been used for testing rocket engines, components and complete stages since 1948. The beginning of the official testing at SSFL was on 15 November 1950 when the first successful main stage test of a US-designed-and-built large liquid propellant rocket engine took place. Because of its stark, remote mountainous appearance the site was used by many Hollywood film makers as the set for films and TV series.

The SSFL facility is currently jointly owned by Boeing (areas 1, 3 and 4) and the NASA Marshall Space Flight Centre (area 2). During the Saturn development ownership was in the hands of NASA and Rocketdyne, who were a division of North American Aviation at the time. The SSFL site consists of four administrative areas used for research, development and test operations and buffer areas on the southern and northwestern boundaries of the facility.

Area 1, to the east, consists of 641 acres owned by Boeing and 42 acres owned by NASA (formerly owned by the US Air Force). Area 1 includes three former rocket engine test areas, two of which were used for the Saturn and Saturn-era programs;

Bowl. As its name suggests this is a natural bowl feature with test positions around the rim. The Bowl was the first test area to be activated and there were originally five test stands built. This area was used for testing Atlas, Navaho, Redstone and Saturn J-2 engines. J-2 engines were fired in Vertical Test Stand 2 (VTS-2), Vertical Test Stand 3A (VTS-3A) and Vertical Test Stand 3B (VTS-3B). J-2 Thrust Chamber Assemblies (TCA) were fired in the Horizontal Test Stand (HTS). Test stand 3A (for altitude) was unique in that a special vacuum diffuser allowed firings to take place at simu-

H-1 engine hot fire test in Canyon Area (1960s)

lated altitude conditions, which were appropriate for the J-2 mission. The original Navaho and Redstone firings were performed in the Vertical Test Stand 1 (VTS-1). Work in the Bowl area has now ceased and most of the infrastructure has been removed.

Canyon. Similar to Bowl, this was another natural feature containing several test stands. Canyon was activated after the Bowl Area, in the mid-to-late 1950s. Canyon has been used to test Jupiter, Thor and Saturn H-1 engines. For H-1 testing there was a two-position stand (horizontal and vertical) and a one-position stand (vertical). The Canyon area is now closed.

In addition, Area 1 contained laboratory facilities where Saturn components were tested;

Component Test Laboratory III (CTL III).
F-1 GG/HX, J-2 Turbo-pumps, J-2 ASI and J-2 GG.

Component Test Laboratory V (CTL V).
J-2 Turbo-pumps.

Area 2, consists of 410 acres in the north-central position of the site and is owned by NASA MSFC. Area 2 contains four test areas, Alfa, Bravo, Coca and Delta, none of which are active today.

Alfa. Used for RS-27 Delta, Atlas, Navaho, Jupiter and Thor engine tests.

Bravo. Used for Atlas, Navaho, Thor, Saturn E-1 (pre-F-1), and Saturn V F-1 components (TCA and Turbo-pump). A gas generator exploded during heat exchanger acceptance testing at Bravo 1C test stand on 7 October 1965.

Coca. Used for Atlas, SSME, S-II-TS-B and Saturn S-II battleship stage testing. The S-II battleship testing was performed at the Coca I site, which was activated in November 1964. Coca IV was constructed for the proposed testing of the S-II-T All-Systems Test stage. However, it was ultimately decided to test that stage at MTF. However Coca IV was used for testing of the structural test stage, S-II-TS-B.

Delta. Used for Atlas, Jupiter, Thor, Saturn IB/V J-2 and Saturn E-1 (pre- F-1). Test stands Delta 2A and Delta 2B were used for J-2 single engine development, qualification and acceptance firings from December 1963.

The Saturn V F-1 component testing at the Bravo test stand ran from 1959-1971.

The Saturn V J-2 testing at the Coca and Delta test

stands took place between 1960 and 1971.

Area 3, consists of 114 acres in the northwest portion of the site and is owned by Boeing. No Saturn-related work was performed in this area.

Area 4, consists of 290 acres owned by Boeing and 90 acres leased by the Department of Energy. The DOE operated several nuclear reactor facilities in this area and in recent years has been the subject of close monitoring to ensure that the ground and surrounding area is free from contamination. It was in this area that a Sodium Reactor Experiment suffered a fuel damage "meltdown" in July 1959.

Buffer zones to the northwest and south consist of 175 and 1,140 acres respectively of undeveloped land.

Boeing acquired Rocketdyne in 1996 and in 2005 Boeing sold the Rocketdyne division to Pratt and Whitney, but SSFL was not part of the transfer. SSFL is no longer active as all future engine testing will be performed at NASA's Stennis Space Centre. In recent years there have been several investigations into the ground contamination in surrounding areas, looking particularly for perchlorate samples. In 2007 Boeing announced that it plans to transfer 2,400 acres to state parkland. As recently as 2008 NASA MSFC conducted a historical survey of the SSFL site.

H-1 being hot fire tested at Canyon 3B Test Stand (1960s)

Pratt & Whitney – West Palm Beach

The Florida Research and Development Center was set up at West Palm Beach in Florida in 1956 by Pratt & Whitney Aircraft, a division of United Aircraft Corporation. The Research Center is located in an isolated 6,750 acre tract on the edge of the Everglades in Palm Beach County.

Following the awarding of the first RL10 LOX/LH2 rocket engine contract in 1958 Pratt & Whitney began constructing the rocket test facilities. All test stands for firing the engines included steam ejectors and diffusers in order to simulate altitude firing conditions for the engines.

During the period of high production in 1962 and 1963 engines were manufactured at Pratt & Whitney's plant at East Hartford, Connecticut, as well as West Palm Beach. The first RL10A-3 engines (the type used on the S-IV stage) to be produced at this plant were completed in July 1962.

The West Palm Beach facility included several test stands for firing the RL10 engines. Horizontal test stands E1 to E4 were used for firing individual engines whilst E5 was a vertical dual engine Centaur Battleship test stand. Test Stands E1 to E4 used a half-shell that sealed around the band on the thrust chamber that was used for supporting the engine during certain handling operations during shipping. During the mid 60s two single engine vertical stands (E6 and E7) were built. These stands were different from the other stands in that they had a continuous vacuum system with a water cooled heat exchanger to cool the exhaust gases before they entered the ejector. The engines were mounted in a capsule and the system provides the capability to operate at variable thrust. E6 is still in use and is used for development activity in addition to production. E6 has a larger diffuser diameter and can test the Atlas version of the nozzle extension. E1 and E2 were later modified to include the capsule arrangement that E6 and E7 have.

Horizontal single engine Test Stand E8 was built on the platform originally used for the E3 and E4 stands. E8 also has a capsule to mount the engine but does not have continuous vacuum.

West Palm Beach continues to be used by Pratt & Whitney Rocketdyne for the production and testing of RL10 rocket engines.

Michoud Assembly Facility (MAF) – Chrysler Corporation Space Division (CCSD)

The 832-acre NASA Michoud Assembly Facility (MAF) is located in New Orleans, Louisiana, some 15 miles east of downtown New Orleans. It is a NASA-owned facility that has been operated by several contractors over the years. During the Apollo program the facility was used by the Chrysler Corporation for building Saturn I/IB stages and by Boeing for building the Saturn V booster stage, the S-IC.

The plant gets its name from Antoine Michoud who operated a sugar cane plantation and refinery in the area in the middle of the 19th century. Two brick smokestacks from the original refinery still stand before the Michoud facility.

The history of Michoud goes back 250 years when a 34,500 acre royal grant of land was obtained from the Governor of the French colony of Louisiana. The original grant was made on 10 March 1763 to Gilbert Antoine de St. Maxent, who was a soldier of the king and established the plantation.

After several ownership transfers the original 34,500 acres had become split amongst various owners. Eventually, in October 1827 Antoine Michoud purchased the remainder of the plantation. Michoud, originally an art dealer, became a successful commercial broker. Michoud died in 1862, his legacy being that he painstakingly bought back all the small lots that had been sold from the main plantation.

The Michoud plantation was later owned by Antoine's nephew, Jean Baptiste Michoud and afterwards by his son Marie Alphonse. The plantation was sold out of the Michoud family in 1910. Various subsequent owners diversified the plantation before the arrival of the Second World War. It was never a very successful sugar plantation and was used in the pre-war years for lumber processing and for the hunting of muskrats.

In 1940 the Government purchased a 1,000 acre tract and constructed the world's largest building at the time, 43 acres under one roof. The plant was originally designed under the direction of the Defense Plant Corporation and construction was initially started with the intention of manufacturing ocean-going vessels. The canals, hydraulic fill and piling for the shipways were constructed, and then prior to start of actual building erection, the plans were changed. The plant was completed on 4 October 1943 as an aircraft factory and used by Higgins Industries of New Orleans for the construction of plywood airplanes. The plywood plane was not satisfactory and the plant was closed on 10 November 1945 after the production of two aircraft.

The plant was inactive between World War II and the start of the Korean Conflict, at which time, in January 1951, it was leased to the Chrysler Corporation for the production of engines for the Sherman and Patton tanks. Chrysler modified the plant for its production needs and installed a complete air-conditioning system. The plant was officially opened on 28 November 1951.

After 1954 the plant was deactivated and maintained by the Government on a standby basis. It was maintained by the US Army Birmingham Ordnance District through a maintenance contract with Machine Products Company of Aberdeen, Mississippi, at a cost of $140,000 per year. Wernher von Braun knew of the site's potential and directed NASA there at the start of the Saturn program.

On 7 September 1961 NASA took over the Michoud Ordnance Plant for the design and manufacture of Saturn first stages. On 10 October 1961 Gurtler Herbert and Co., Inc. of New Orleans, was awarded a contract for re-habitation and modification of the facility. The first Saturn stage built by Chrysler at Michoud was handed over to NASA at a ceremony on 13 December 1963. The facility includes a port with deep-water access for the transportation of large space structures. The facility also includes a high bay area, the Vertical Assembly Building, where Saturn V S-IC stages were assembled vertically. The VAB is a single story structure rising the equivalent of 18 stories, and was completed on 15 December 1964.

Checkout of the S-I/IB and Saturn V stage's electrical and mechanical systems was performed in the four giant test cells of the Stage Test Building. Each of the Cells is 83 by 191 feet with 51 feet of clear height. Each had separate test and checkout equipment.

Stages left and entered Michoud by waterways connected to the Mississippi River or the Gulf of Mexico. On 8 June 1967 the facility changed its name from Michoud Operations to the Michoud Assembly Facility.

In 1973, with the Apollo program winding down Chrysler and Boeing made way for Lockheed Martin Michoud Space Systems who designed and manufacture the Space Shuttle External tanks at Michoud. The first tank rolled off the Michoud production line on 9 September 1977. The same barges as used to transport the Saturn rockets are still used today to move the External tanks. The facility will be used in the future to build the Ares launch vehicles.

S-I stage simulator outside Michoud entrance (Late 1963)

The Douglas Aircraft Company (DAC), Missile and Space Systems Division - Santa Monica

Santa Monica on the Los Angeles coast was the Douglas facility used for assembly of the S-IV stage. S-IV tank panels were formed at the Douglas plant at El Segundo, Los Angeles. The inter-stage panels were formed at the Douglas plant at Long Beach. The panels were shipped to Santa Monica where the characteristic waffle pattern was milled onto the inside surface.

Douglas Santa Monica plant and Santa Monica Airport (Early 1960s)

At Santa Monica there were various production areas used in manufacturing the S-IV stages. Production Area 136 was the major assembly build-up area for the forward, aft and common bulkheads. In addition, x-ray inspection of these assemblies was conducted in this area. Production Area 46 was used for welding cylindrical tank segments. The Pandjiris automatic seam welder, used for longitudinal welding of tank sections to form the cylindrical assembly, was located in this area.

Production Area 17 was used for panel pressure checks and subassembly installation of the forward inter-stage; build up of heat shields, and thrust structure and umbilical panels; and assembly of the forward inter-stage, aft inter-stage and skirt panels into completed sections. In addition this area was used for installation checkout, inspection of the six RL10 engines; painting, weighing and final loading of the stage on the transporter.

The Etch and Bonding Area was used for all common bulkhead etch and bonding.

The Assembly and Calibration Test Tower was used for assembly of the S-IV vehicle and for leak testing and calibration of the liquid oxygen tanks.

In 1965 the Assembly Towers were transferred to the Huntington Beach facility of DAC where the S-IVB-200 stages were assembled. Later the Santa Monica plant was demolished and now no longer exists.

S-IV Assembly and Checkout Towers at Santa Monica
(Early 1960s)

The Douglas Aircraft Company (DAC), Missile and Space Systems Division - Huntington Beach

The Douglas Aircraft Company, based at Huntington Beach in southern Los Angeles, designed and manufactured the S-IVB-200 second stage at that facility. The center was the headquarters for the Missile and Space Systems Division. Construction of the Huntington Beach facility started in early January 1963. During May 1963 the fabrication and assembly building was completed and construction of the assembly towers began. Additionally, in 1965 the former S-IV assembly towers were transferred from DAC's Santa Monica plant.

Initial component fabrication for the second stage was accomplished at Douglas' Santa Monica plant. Final assembly and factory checkout of the second stage took place at the Space Systems Center at Huntington Beach, just down the coast from Santa Monica. High bay manufacturing area was provided for the production of propellant tanks, skirts and inter-stages. Eight tower positions were available for vertical assembly and checkout of completed vehicles. Structural tests on major vehicle structures such as the propellant tank, skirt sections, and inter-stage were conducted in the

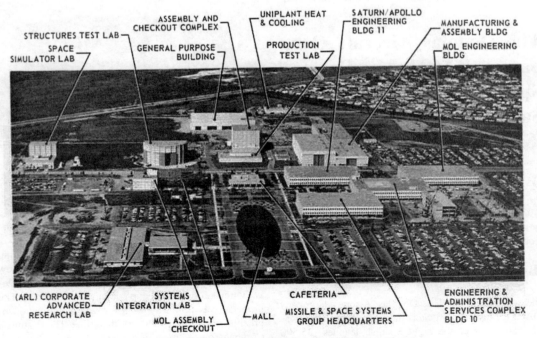

Douglas Missile & Space Systems Division, Huntington Beach (1966)

Structural Test Laboratory at the Space Systems Center.

Two vertical towers at Huntington Beach provided for the final factory tests on finished second stages, prior to shipment from the plant for test firing. Completed stages were transferred by road to the nearby Los Alamitos Naval Air Station and airlifted on a Super Guppy aircraft to Douglas' Sacramento Test Facility for static firing. The stages were then transported directly from Sacramento to KSC.

In April 1967 Douglas became McDonnell Douglas and the Huntington Beach plant is now owned by Boeing.

Entrance to Beta complex (1965)

Assembly and Checkout Towers, Huntington Beach (1960s)

Entrance to Beta complex - present day (2006)

The Douglas Aircraft Company (DAC), Missile and Space Systems Division - Sacramento Test Operations (SACTO)

SACTO road signs (2006)

Overview

SACTO was a test facility used by the Douglas Company to checkout and test all the S-IV and S-IVB stages produced for the Saturn program.

History

The area used by The Douglas Aircraft Company (DAC) for test firing rockets throughout the 1960s was originally part of the eight Spanish leagues (35,500 acre) Rancho Rio de los Americanos land granted to

William Leidesdorff by the Mexican government when he applied for Mexican citizenship in 1844. Joseph L Folsom purchased the rights to this land, 15 miles east of Sacramento, in 1848 following Leidesdorff's death.

Agriculture was the main industry in the region during the late 19th and early 20th centuries, and the area now known as Rancho Cordova was named after the Cordova Vineyard, which occupied part of that area. The property was primarily used for wheat cultivation or grazing until the 1920s. At that time The Natomas Company owned most of the land. After the discovery of gold in the area extensive dredging took place from

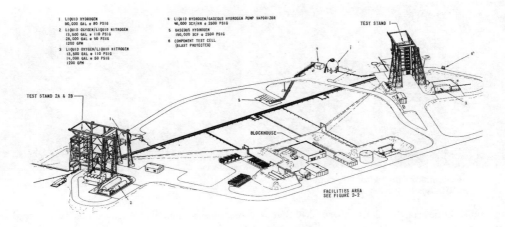

1 LIQUID HYDROGEN
  90,000 GAL @ 80 PSIG
2 LIQUID OXYGEN/LIQUID NITROGEN
  13,500 GAL @ 110 PSIG
  28,000 GAL @ 50 PSIG
  1200 GPM
3 LIQUID OXYGEN/LIQUID NITROGEN
  13,500 GAL @ 110 PSIG
  14,000 GAL @ 50 PSIG
  1200 GPM
4 LIQUID HYDROGEN/GASEOUS HYDROGEN PUMP VAPORIZOR
  40,000 SCF/HR @ 2500 PSIG
5 GASEOUS HYDROGEN
  150,000 SCF @ 2500 PSIG
6 COMPONENT TEST CELL
  (BLAST PROTECTED)

TEST STAND 1

TEST STAND 2A & 2B

BLOCKHOUSE

FACILITIES AREA
SEE FIGURE 2-2

Alpha complex (1966)

1915 to 1962. It has been estimated that The Natomas Company removed $100 million of gold during this period. What was left behind were the characteristic dredging piles – long parallel mounds of rock that rise above the landscape.

Beginning in 1950, Aerojet began purchasing parcels of dredged land from The Natomas Company. This barren land was completely unusable for most exploitation, but perfect for establishing a remote rocket testing facility. By the mid 1950s Aerojet had purchased around 25,000 acres 15 miles east of Sacramento.

The Douglas Aircraft Company entered into a lease agreement with Aerojet in 1956 to take over approximately 1,700 acres on the western edge of the Aerojet site in order to provide a remote test site to test Thor missiles. Douglas subsequently purchased 3,800 acres from Aerojet in 1961, which included the previously leased 1,700 acres. This site was known as the Sacramento Test Operations or SACTO.

In 1956 construction at the Douglas test site started. Initially a single position vertical missile test stand (TS1)

and one dual position vertical missile test stand (TS2A and TS2B) were constructed. These stands, which later became known as the Alpha 1 and Alpha 2A/B stands, were completed in 1957.

With the advent of the Saturn family of launchers Douglas was awarded a number of contracts. Initially they were awarded a contract on 26 April 1960 to design, build, and test the S-IV second stage for the Saturn I booster. Up to that point Douglas had considerable experience with LOX but none with LH2.

By the summer of 1961, in preparation for activities with the S-IV stage, the modifications to the TS1 test stand to accommodate the 18.5 feet diameter S-IV stage were completed. Test stand TS2B was modified for S-IV testing and conversion was completed in January 1963. Estimated cost of the modifications at the time was $2 million.

At the time there were great unknowns about the use of gaseous and liquid hydrogen. It was believed that any leaks of hydrogen would easily combust and because the flame emissivity was low it was feared that the

Alpha complex (Early 1960s)

Installing the LH2 tank at Alpha 2A/B site (1962)

Alpha complex with TS1 on the right and TS2A/B on the left (Early 1960s)

Present day remnants of Alpha TS1 (2006)

flames would not be seen. So, during these early Battleship test activities one of the last events before securing the test stand was for the test crew to "walk around" the loaded vehicle and look for leaks or other anomalies.

Without previously existing procedures the engineers at SACTO were writing the handbook as they went. They decided that a simple means to detect an otherwise invisible on-going fire was to wave a flammable material in the suspect area. They contrived to use straw brooms, as they were cheap, readily available, and easily ignited. The "walk around" crew waved brooms into each of the areas that were considered capable of leaking such as the engine compartment, propellant feed line joints, and GH2 gas line joints. For the early Battleship firings the Douglas technicians attached strings

of twine around various locations of the engine and watched these on video to see if they caught fire. For later firings GE Infrared cameras were installed which could detect low emissivity hydrogen flames.

Following the award of the contract for the S-IVB stage to be used on both the Saturn IB and Saturn V launch vehicles a new test complex was designed and constructed. This became known as the Beta complex and comprised two test stands, Beta I and Beta III.

Following the final test firing of stage S-IVB-511 on 18 December 1969 the test stands and supporting infrastructure were moth balled by McDonnell Douglas (the successor to DAC). It was anticipated that NASA would restart production activities at a later date. How-

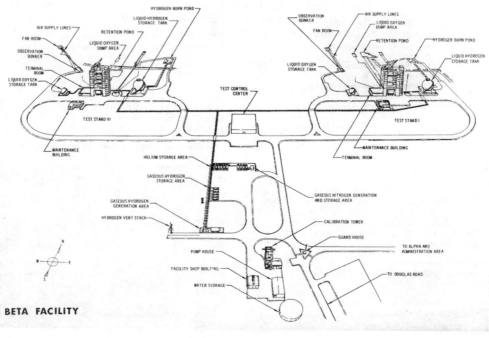

Beta complex (1966)

**Beta complex with Beta I on the right
and Beta III on the left** (1965)

Remnants of Beta I test stand (2006)

ever, this never materialized and the resources were relocated to other jobs.

Meanwhile NASA started removing all the high value equipment from the site such as tape recorders, computers, and electronic consoles. A Japanese company removed the steel superstructures from the Alpha and Beta test stands in the late 1970s for a nominal sum. What remains of the actual test stands is a large concrete plinth at each stand.

Until 1977 the site was inactive. From 1977 to 1984 McDonnell Douglas leased or sold portions of the Administration Area as a venture park for small businesses and renamed the area Security Park. The old VCL building still bears the name "Security Park" on its side and is easily visible from Douglas Road.

In 1984 McDonnell Douglas negotiated with Aerojet for the return of the 3,800-acre area to Aerojet ownership. The exception was the former Administration Area which continued to be owned and used by various small businesses. For the past 15 years Aerojet has exhaustively inspected and tested the environmental condition of the soil and groundwater in preparation for subdividing the land for commercial businesses and private dwellings. The area of Aerojet property that was once owned by Douglas is now known as the Inactive Rancho Cordova Test Site (IRCTS).

Description of facilities

General

The Douglas site in Sacramento was variously known as Sacramento Test Operations (SACTO), and Sacramento Test Centre (STC). It was operated as an out-station of the main manufacturing plants in Santa Monica and later Huntington Beach. Douglas used a numbering system for their many locations, e.g. "A" related to Santa Monica (space), "B" related to El Segundo (Naval aircraft), and "C" related to Long Beach (Air Force aircraft). Within the space category A45 covered SACTO, A41 was for Cape Canaveral/Kennedy Space Centre, and A31 was for the White Sands Proving Grounds.

Alpha site

The Alpha complex was the first built at SACTO. It was located approximately one mile from the Administration and Support area and occupied an area of about 45 acres. It included two test stands (3 test positions) identified as TS1 and TS2A/B. The site included one 90,000 gallons liquid hydrogen tank at each of the two test stands, which were pressurized at 80 psig. Each test stand also had two liquid oxygen storage vessels pro-

Beta I test stand (1965)

Beta complex control room (Mid 1960s)

Remnants of Beta complex control room (2006)

Remnants of Gamma complex (2006)

tected by a high dirt/concrete barricade around them. At the TS1 these vessels held 13,500 gallons of LOX at 110 psig and 14,000 gallons of LOX at 50 psig respectively. At TS2A/B these vessels held 13,500 gallons of LOX at 110 psig and 28,000 gallons of LOX at 50 psig respectively. Originally there were facilities for the supply of RP-1 kerosene.

Test Stand 1 was a steel beam structure on a concrete base and foundation. The stand was designed for a maximum thrust level of 300,000 pounds and was used for both the Thor and S-IV vehicles. It was 700 feet from the control centre and 1,200 feet from Test Stand 2. The stand included a 10-ton capacity overhead crane

Warning light box in Beta control room (2006)

for lifting the vehicles onto the stand.

Test Stand 2 was a dual position stand of steel structure on a concrete base and foundation. Both adjacent positions could be utilized at the same time. The stand was designed for thrust levels up to 300,000 pounds. Stand 2A was utilized for the Thor missile and for Titan engines whilst Stand 2B was used for the Thor and S-IV stages. It too had a 10-ton capacity crane on the stand capable of handling vehicles on both the 2A and 2B positions.

An altitude simulation system was built for both the TS1 and TS2B test stands comprising four elements: the exhaust diffusers, the ejectors, the accumulators and the steam boilers and water feed system. The diffusers were attached to each of the six RL-10 engines with a flexible seal and were closed at the opposite end with a blow-off door. In this configuration they served as a vacuum chamber providing a pressure of less than 0.9 psia during the 45 seconds prior to engine ignition. By controlling the engine exhaust gas flow through internal

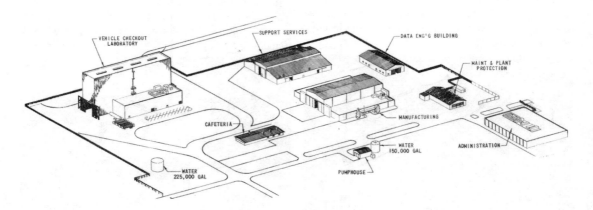

VCL and administrative area (1966)

Administrative area (Early 1960s)

geometry the diffusers were also able to sustain the required absolute pressure at the engine bell exit plane after the engine start transient. The diffusers were approximately 35 feet long and were of double wall construction to provide for water cooling.

Each diffuser was connected to a two-stage steam ejector. Each stage of the ejector was 30 feet long and they were assembled together in a vertical array on the front of the test stand. Two 30,000 gallons steam accumulators served as storage vessels for the steam used to power the ejectors. A boiler was used to produce 8,625 pounds per hour of dry and saturated steam at 250 psia for charging the steam accumulators. The process of charging the accumulators required about 12 hours.

After successful engine start the steam ejector system was isolated thus relying on the engine exhaust to maintain a pressure of less than 1.0 psia at the nozzle exit plane. The diffuser blow-off doors, used to maintain a vacuum in the diffuser during steam ejector operation, were blown open at engine start. During the initial static firings of the S-IV Battleship aluminum doors were used at the closure of the diffusers. These weighed 120 pounds and were held in place by large magnets. As they blew off they tended to be damaged and experience showed that they could not be used more than two times. To overcome this problem lighter (53 pound), more flexible, fiberglass doors were installed for later firings.

Administrative building (Mid 1960s)

The Test Control Centre was a reinforced concrete building protected by earth barricades. The control centre contained individual firing and facility consoles for all three test positions of the test stands. There was a capability of recording 750 channels of data as well as 11 closed loop television circuits.

Beta site

On 21 December 1961 NASA awarded Douglas a contract to design, build, and test the upgraded S-IVB stage to be used in two configurations. The 200 model was used as the second stage of the Saturn IB launcher and the 500 model was used as the third stage of the Saturn V launcher.

To support this new program Douglas initiated work on a new test complex that became known as the Beta complex. TS1 and TS2A/B accordingly were renamed the Alpha complex. As NASA had seemingly unlimited funds at that time they instructed Douglas to design for three Beta test stands with a central control room blockhouse.

Following direction from NASA to proceed with the Beta complex in August 1962 The Ralph M Parson Company completed the design of the Beta test stands on 6 November 1962. Acoustical considerations indicated that the site should be positioned 17 degrees east of north.

Construction of the Beta complex started on 26 November 1962 when ground breaking occurred. By February 1963 the preliminary road layout and site excavation had been completed with the moving of approximately 420,000 cubic yards of earth. The pile driving activity was completed one day ahead of schedule on 15 March 1963. This involved driving 855 piles in lengths of 100, 120, and 140 feet for a total of approximately 92,000 feet. Communications cabling from Alpha to Beta sites was completed on 18 March 1963 in one week, despite unfavorable weather. The prime construction contract was awarded to Paul Hardeman Company of Los Angeles who started to install equipment on site on 25 March 1963. Construction of the complex required approximately 9,200 cubic yards of concrete. The large liquid hydrogen tanks were constructed by the Chicago Bridge and Iron Company.

The Beta I stand was activated in September 1964 with the first cryogenic loading in the S-IVB Battleship stage. Meanwhile the second Beta test stand, Beta III, became operational and was used for the first flight stage firing of S-IVB-201 on 8 August 1965. Between 1965 and 1969 nine S-IVB-200 and eleven S-IVB-500 flight stages were static fired at SACTO.

The Beta complex was located two miles west of the Administration and Support area. The Beta complex proper occupied approximately 325 acres. A further 3,800 feet buffer zone contained approximately 1,125 acres. The buffer zone was calculated on the basis that the maximum loading of LOX and LH2 was equivalent to a weight of 150,000 pounds of TNT explosive which, if detonated, would produce an overpressure wave no greater than 0.5 psig at the nearest road, Sunrise Blvd. The test stand foundations and structures were designed for a thrust capability of 1.0 million pounds although only one quarter of this level was ever utilized.

The test control centre, or blockhouse, was a steel-reinforced concrete structure with a usable interior area of 26,240 square feet. The centre contained parallel services for the Beta I and Beta III test stands, including all Facility Control Systems, Data Acquisition Systems, and closed-circuit television receivers. There were 868 channels of recording capability in the test center. The rooms were fully air-conditioned and they were joined to the test stands by underground tunnels. The control center was located 1,000 feet from each test stand.

The calibration tower, at the entrance to the complex, was a structure designed for performing precise calibration of the vehicle cryogenic weighing system. The pump house was a steel-reinforced concrete structure containing four 10,000 GPM water pumps. A diesel-driven generator was also installed to provide emergency electrical power. A 1.5 million gallons cylindrical steel water tank was located next to the pump house. A high-pressure gas storage and generation area included facilities for the storage and supply of helium, hydrogen, and nitrogen.

Although originally designed to support three test stands only two were finally constructed. The Beta I and III test stands were identical and stood 150 feet high. They were 1,800 feet apart. Each stand comprised a concrete pedestal supporting steel superstructure inclusive with work platforms for accessing the rockets

Present day view of administrative building (2006)

at various heights. A 15-ton capacity jib-crane at the ninth deck level and a 50-ton capacity boom-crane at the thirteenth level were installed for vehicle/stage handling and for weighing operations. A shop building was located adjacent to each stand to provide mechanical support, and to provide air supply requirements. A terminal room, which contained ground support equipment and control interfaces, was attached to the side of each test stand.

During firings a 40 feet high deflector plate was located at the engine exhaust and was cooled by a water flow of approximately 20,000 GPM through 3,300 5/32 inch diameter holes. Each stand was supported by one liquid hydrogen and one liquid oxygen tank. Each tank was double-walled with a Pearlite insulation material between the skins. The liquid hydrogen tanks were 41 feet in outer diameter and had a capacity of 175,000 gallons. The liquid oxygen tanks were 27 feet in outer diameter and had a capacity of 40,000 gallons.

A downrange observation bunker was located 260 feet from each stand and allowed good visibility of the test firings through reinforced glass windows.

Gamma site

The contractor for the construction of the Gamma complex was Wismer-Becker who completed work on 22 July 1964. Douglas, who activated the site two months later, used it for hot fire testing the bipropellant Auxiliary Propulsion Modules used for attitude control and ullage settling on the S-IVB stages.

The Gamma complex test area was located 0.5 miles northwest of the Administration and Support area. The Gamma complex test area consisted of a test structure containing three test cells; test control center, instrumentation center, maintenance and assembly building, and storage areas for the propellants. The main propellants that were used here were monomethyl hydrazine fuel and nitrogen tetroxide oxidizer. The cells were identified as Cell I, Cell II, and Cell III. The cells were separated by concrete walls and covered with a removable roof. Each test cell was 18-foot square.

Test Cell III was configured to test either module or cluster type hypergolic engines. Fuel and oxidizer for each test cell was supplied from a mobile servicer. Located 50 feet from the test cells was the test control center, built of a reinforced concrete structure. A separate instrumentation center was located 350 feet from the test control center.

Kappa site

The Kappa site was located 0.5 miles northwest of the

Administration and Support area and adjacent to the Gamma site. This area was used for component and subsystem development and production acceptance testing. Test Cells A, B, C, and E were used for testing involving liquid hydrogen and helium gas.

Vehicle Checkout Laboratory and Administration and Support area

Site preparation for the Vehicle Checkout Laboratory (VCL) was started in October 1964. This included ground leveling, grading and surfacing. Actual construction of the building commenced in December 1964.

Located at the entrance to SACTO, immediately off Douglas Road, was the Administration and Support Area, occupying about 70 acres. Originally this comprised administration, manufacturing, support services, maintenance & plant protection, data engineering, and cafeteria buildings. Later, to serve the S-IVB stage, the huge Vehicle Checkout Laboratory was constructed in the same area.

Processing of the S-IV stages at SACTO occurred prior to the construction of the VCL. For those stages the pre- and post-firing testing took place in the Support Services Building, which is still standing, and located close to the VCL. The S-IV stages were held on horizontal transporters in the central high bay of this building. The data laboratory was adjacent to the high bay and allowed for umbilical connection between the stage and the IBM 360 computer which played the checkout tapes and recorded system performance.

The VCL was designed for performing simulated flights and conducting post fire checks on the S-IVB stages. The VCL comprised two towers (north and south), a control room, and maintenance support areas. The two towers situated side by side were separated by a blast wall. Several S-IVB stages could be accommodated in each tower at any one time, some vertically and some horizontally. The vertical stages were supported by a number of horizontal work platforms that rotated into place. The control room was located on the second floor of the VCL and contained the necessary control consoles and equipment to monitor the health of the stages. The VCL had 182 channels of instrumentation recording capability.

There was a metrology laboratory, machine shop and an electrical laboratory located within the Manufacturing Building.

Current status of the facility

The Beta I and III test stands remain except for the superstructure. The water-cooled flame deflector plates have been removed but their concrete supports remain in place, redundant for the past 40 years. Two of the four propellant storage tanks have been removed. The liquid hydrogen tank at Beta I and the liquid oxygen tank at Beta III remain although they have had their tops cut off.

The Beta control rooms' false floors and ceilings have been removed and copper strips have been stolen over the years. However, one reminder of the past remains in place to safeguard any future rocket testers. Hanging from the roof there is a warning panel with three boxes that in days gone by would shine brightly green or red. The words spell out the dangers – man in engine – man in forward skirt – man in burn pond. The panel was installed following an accident on the S-IV battleship stage that was being prepared on the Alpha TS1 stand. A technician climbed up inside the nozzle to perform an inspection of an engine injector for erosion and cleanliness. The ground crew, not realizing he was there, turned on a nitrogen trickle purge, needed to maintain cleanliness. The man lost consciousness and fell to his death at the bottom of the diffuser, striking his head on the steel work platform.

The Alpha test stands have been stripped of their steelwork and anything of value. However the concrete pedestals and flame trenches remain in good shape. Unlike the Beta stands these were built on the side of a natural tailings mound to improve the dispersion of the exhaust efflux.

The Gamma facility also is cordoned off but appears to look exactly as it did in the 1960s, ready to support firings of the bipropellant Auxiliary Propulsion Units.

Since about 2000 the main administration building and the VCL have been occupied by Automotive Importing Mfg., Inc, who use the VCL to store automotive parts prior to onward delivery. The parts are stacked on the VCL floor up to a height of some 10 feet leaving the remainder of the north and south bays of the tall VCL unused and looking exactly as they did when they serviced S-IVB stages held vertically.

The future development of SACTO depends largely on the success of the land decontamination process and the ability of the Aerojet Real Estate Division to market the area. In any event the test stand concrete pedestals are likely to be very costly to remove and so may remain for some years to come as a totally unrecognized monument to a great era of exploration.

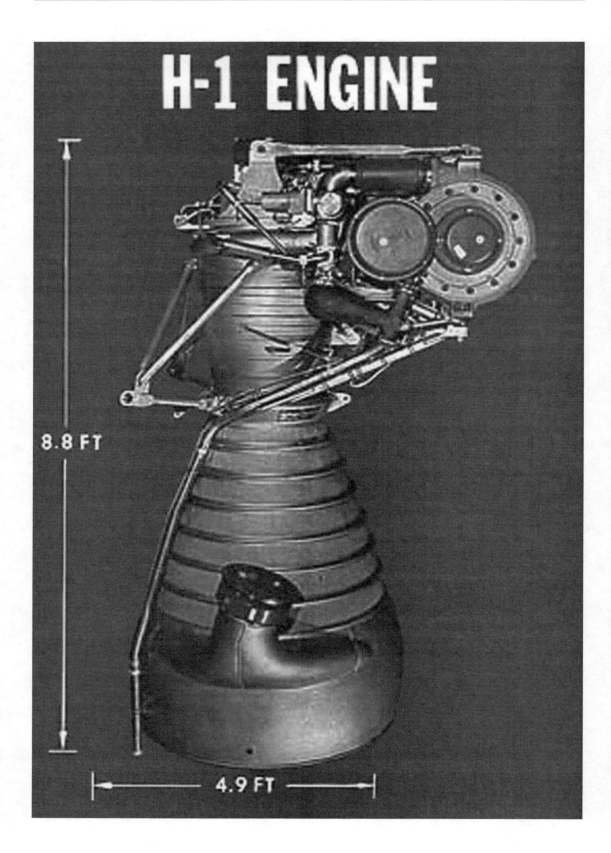

H-1 ENGINE

8.8 FT

4.9 FT

H-1 Engine

The Rocketdyne Division of North American Aviation was awarded a contract to design, manufacture and test the H-1 rocket engine on 11 September 1958. The H-1 engine was an outgrowth of the combined knowledge gained from the Redstone, Thor, Jupiter and Atlas engines. Design and initial manufacturing took place at Rocketdyne's Canoga Park facility. Engine development and qualification testing was performed in the Canyon Area at the Santa Susana Field Laboratory. The first engine main-stage test occurred on 31 December 1958 and the first prototype engine, H-1001, was first tested on 6 March 1959 and was dispatched from Rocketdyne to ABMA on 20 April 1959.

MSFC Power Plant Test Stand for single H-1 engine testing (Late 1950s)

The initial design thrust level was 188,000 pounds, but the first group of engines was down-rated to 165,000 pounds thrust.

The first flight engine was acceptance tested on 14 October 1959 and was delivered to ABMA in Huntsville on 27 January 1960. On 1 November that year the formal 165,000 pounds thrust engine PFRT program was completed using engines H-1031 and H-1037. A total of 34 tests with 2,507 seconds of main-stage duration were achieved between 16 September and 1 November 1960. Engines H-1001 to H-1060 were at the 165,000 pounds thrust level. Saturn I Block I vehicles SA-1 to SA-4 utilized engines at this thrust level.

The first baffled-injector and 188,000 pounds thrust engine (H-5001) started testing on 9 January 1961 and was delivered on 12 February 1962. The formal 188,000 pounds thrust PFRT program was completed on 28 September 1962, using engine H-5018. Engines H-5001 to H-5045 were at the 188,000 pounds thrust

level. Saturn I Block II vehicles SA-5 to SA-10 utilized engines at this thrust level.

The first 200,000 pounds thrust engine (H-4044) was delivered on 31 March 1964. The engine-level qualification program for the 200,000 pounds thrust engine was completed on 7 April 1965, using engines H-4055 and H-7055. The qualification program had run from 10 March to 7 April during which each of the two engines had been tested in excess of the qualification life of 17 tests or 2,325 seconds. The component qualification program for the 200,000 pounds thrust engine was completed on 27 April 1965. Engines H-4044 to H-7070 were at the 200,000 pounds thrust level. Saturn IB vehicles AS-201 to AS-205 utilized engines at this thrust level.

The first 205,000 pounds thrust level engine (H-4068) was delivered on 21 October 1965. The engine re-qualification program at the 205,000 pounds thrust level

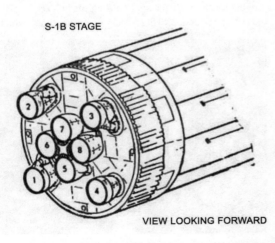

S-1B STAGE

VIEW LOOKING FORWARD

H-1D OUTBOARD ENGINE

POSITIONS 1, 2, 3, & 4
7XXX SERIAL NUMBERS

* 306,000 LB THRUST EACH
* BIPROPELLANT: LOX/KEROSENE
* GIMBALED THRUST VECTOR
* ASPIRATED TURBINE EXHAUST

H-1C INBOARD ENGINE

POSITIONS 5, 6, 7, & 8
4XXX SERIAL NUMBERS

* 306,000 LB THRUST EACH
* BIPROPELLANT: LOX/KEROSENE
* FIXED THRUST VECTOR
* DUCTED TURBINE EXHAUST

H-1 engine location in S-I/IB stages

H-1 engine being fired at the MSFC Power Plant Test Stand (1960s) 6778743

was completed on 16 June 1966, using engine H-4083. Engines H-4068 to H-7119 were at the 205,000 pounds thrust level. Saturn IB vehicles AS-206 to AS-210, as well as stages S-IB-11, -12, -13 and -14 utilized engines at this thrust level.

Rocketdyne completed testing of the first flight worthiness verification engine on 20 October 1966. This program verified the integrity of a typical flight engine after long term storage in the field. They also completed the 205,000 pounds thrust engine component qualification program on 26 May 1967.

Although development, qualification and early flight production manufacturing and testing took place at Canoga Park and Santa Susana respectively, manufacturing and flight acceptance testing soon transitioned to another Rocketdyne plant at Neosho, Missouri, formerly Air Force Plant No 65. The first H-1 engine test at Neosho was on 25 May 1962. The first H-1 production

engine to be delivered from the Rocketdyne plant at Neosho occurred in September 1962. Prior to this two Reliability Verification Engines (RVT) were fabricated at Neosho.

By January 1968 production of further H-1 engines was terminated. Consequently, in January 1968 NASA and Rocketdyne decided to move the site of H-1 engine support and maintenance production from Neosho, Missouri, back to Canoga Park, California. Also, Rocketdyne re-activated the Canyon 3B Test Stand at SSFL in April 1968 to accommodate H-1 testing following the transfer. The first engine assembly and test back in Los Angeles was achieved in June 1968.

Rocketdyne successfully completed the H-1 engine flight verification program in July 1968. The test program involved engine H-7053, the flight worthiness engine for the S-IB-5 stage. This engine had accumulated 17 starts and 2,116 seconds of firing time, which

H-1C inboard engine H-4077, later to fly on AS-208
(1.3.1966)

H-1C inboard engine H-4077, later to fly on AS-208
(1.3.1966)

H-1D outboard engine (25.2.1966)

H-1D outboard engine (25.2.1966)

exceeded the minimum qualification life requirement of 15 starts and 2,025 seconds.

During the course of the H-1 program a number of problems were encountered. These can be classified into the groups, combustion instability, LOX dome cracks, thrust chamber tube splits, thrust OK pressure switch (TOPS) vibration problems, LOX pump seal problems, heat exchanger coil leaks, turbine blade failures and LOX pump shaft seal problems. There was one failure in flight, when engine H-2007 shut down prematurely during the launch of SA-6.

On Saturn stages the inboard engines were designated H-1C and the outboard engines H-1D. In late 1971 development of the RS-27 engine for the Delta launch vehicle was initiated at Rocketdyne. The engine was a Thor derivative with H-1 components packaged in.

With the conclusion of the Saturn I and IB programs there was a surplus of H-1 engines. Some were transferred to the Atlas program, but most were transferred to the RS-27 engine program for use on the Delta launch vehicle. The similarity between the H-1 and RS-27 resulted in the spare H-1 engines being cannibalized for parts for the RS-27 engines. In general the RS-27

utilized turbo-pumps, turbines, gas generators, valves and thrust chambers from the H-1 engines.

RS-27 engine production continued until 1983 and many of the engines produced in this time utilized former H-1 engine components. Following the Challenger explosion the RS-27 production line was re-started in 1987, but this time the engines were built from scratch without the use of former H-1 engine parts.

H-1 engine being fired at Neosho (1960s) 6418296

H-1D outboard engine (25.2.1966)

Listing of all H-1 engines

MSFC SN H-1001. Rocketdyne SN H-1001
The first prototype engine dispatched from Rocketdyne on 20 April 1959. The first firing of this engine in the Power Plant Test Stand at ABMA in Huntsville occurred on 26 May 1959. The engine is now installed on S-I-T at MSFC.

MSFC SN H-1002. Rocketdyne SN H-1002
The engine is installed on S-I-T at MSFC.

MSFC SN H-1003. Rocketdyne SN H-1003
The engine is installed on S-I-T at MSFC.

MSFC SN H-1004. Rocketdyne SN H-1004
Test Division spare, scrapped.

MSFC SN H-1005. Rocketdyne SN H-1005
The engine is installed on S-I-T at MSFC.

MSFC SN H-1006. Rocketdyne SN H-1006
The engine is installed on S-I-T at MSFC.

MSFC SN H-1007. Rocketdyne SN H-1007
The engine is installed on S-I-T at MSFC.

MSFC SN H-1008. Rocketdyne SN H-1008
The engine is installed on S-I-T at MSFC.

MSFC SN H-1009. Rocketdyne SN H-1009
The engine is installed on S-I-T at MSFC.

MSFC SN H-1010. Rocketdyne SN H-1010
Test Division spare, scrapped.

MSFC SN H-1011. Rocketdyne SN H-1011
The engine was installed in position 105 on S-I-1 and launched on SA-1 on 27 October 1961.

H-1 engines awaiting stage installation at MAF (29.8.1967)

MSFC SN H-1012. Rocketdyne SN H-1012
The engine was installed in position 106 on S-I-1 and launched on SA-1 on 27 October 1961.

MSFC SN H-1013. Rocketdyne SN H-1013
The engine was installed in position 107 on S-I-1 and launched on SA-1 on 27 October 1961.

MSFC SN H-1014. Rocketdyne SN H-1014
Test Division spare, scrapped.

MSFC SN H-1015. Rocketdyne SN H-1015
The engine was installed in position 108 on S-I-1 and launched on SA-1 on 27 October 1961.

MSFC SN H-1016. Rocketdyne SN H-1016
The engine was installed in position 101 on S-I-1 and launched on SA-1 on 27 October 1961.

MSFC SN H-1017. Rocketdyne SN H-1017
The engine was installed in position 102 on S-I-1 and launched on SA-1 on 27 October 1961.

MSFC SN H-1018. Rocketdyne SN H-1018
The engine is installed on S-I-D at MSFC.

MSFC SN H-1019. Rocketdyne SN H-1019
The engine was installed in position 103 on S-I-1 and launched on SA-1 on 27 October 1961.

MSFC SN H-1020. Rocketdyne SN H-1020
Test Division spare, scrapped.

MSFC SN H-1021. Rocketdyne SN H-1021
The engine was installed in position 104 on S-I-1 and launched on SA-1 on 27 October 1961.

MSFC SN H-1022. Rocketdyne SN H-1022
Spare for SA-1 and SA-2, scrapped.

MSFC SN H-1023. Rocketdyne SN H-1023
The engine is installed on S-I-D at MSFC.

MSFC SN H-1024. Rocketdyne SN H-1024
MSFC Orientation Center.

MSFC SN H-1025. Rocketdyne SN H-1025
The engine is installed on S-I-D at MSFC.

MSFC SN H-1026. Rocketdyne SN H-1026
The engine is installed on S-I-D at MSFC.

MSFC SN H-1027. Rocketdyne SN H-1027
On permanent loan to University of South Dakota.

MSFC SN H-1028. Rocketdyne SN H-1028
The engine was installed in position 105 on S-I-2 and launched on SA-2 on 25 April 1962.

MSFC SN H-1029. Rocketdyne SN H-1029
The engine was installed in position 106 on S-I-2 and launched on SA-2 on 25 April 1962.

MSFC SN H-1030. Rocketdyne SN H-1030
Test Division spare, scrapped.

MSFC SN H-1031. Rocketdyne SN H-1031
165,000 pounds thrust PFRT. Disassembled and scrapped.

MSFC SN H-1032. Rocketdyne SN H-1032
The engine was installed in position 101 on S-I-2 and launched on SA-2 on 25 April 1962.

MSFC SN H-1033. Rocketdyne SN H-1033
The engine was installed in position 102 on S-I-2 and launched on SA-2 on 25 April 1962.

MSFC SN H-1034. Rocketdyne SN H-1034
The engine was installed in position 103 on S-I-2 and launched on SA-2 on 25 April 1962.

MSFC SN H-1035. Rocketdyne SN H-1035
The engine was installed in position 107 on S-I-2 and launched on SA-2 on 25 April 1962.

MSFC SN H-1036. Rocketdyne SN H-1036
The engine was installed in position 108 on S-I-2 and launched on SA-2 on 25 April 1962.

MSFC SN H-1037. Rocketdyne SN H-1037
165,000 pounds thrust PFRT. Disassembled and scrapped.

MSFC SN H-1038. Rocketdyne SN H-1038
The engine was installed in position 104 on S-I-2 and launched on SA-2 on 25 April 1962.

MSFC SN H-1039. Rocketdyne SN H-1039
Spare, stability test engine, scrapped.

MSFC SN H-1040. Rocketdyne SN H-1040
The engine was installed in position 105 on S-I-3 and launched on SA-3 on 16 November 1962.

MSFC SN H-1041. Rocketdyne SN H-1041
The engine was installed in position 106 on S-I-3 and launched on SA-3 on 16 November 1962.

MSFC SN H-1042. Rocketdyne SN H-1042
The engine was installed in position 107 on S-I-3 and launched on SA-3 on 16 November 1962.

MSFC SN H-1043. Rocketdyne SN H-1043
The engine was installed in position 108 on S-I-3 and launched on SA-3 on 16 November 1962.

MSFC SN H-1044. Rocketdyne SN H-1044
Spare for SA-3. On permanent loan to University of Mississippi.

MSFC SN H-1045. Rocketdyne SN H-1045
The engine was installed in position 101 on S-I-3 and launched on SA-3 on 16 November 1962.

MSFC SN H-1046. Rocketdyne SN H-1046
The engine is installed on S-I-D at MSFC.

MSFC SN H-1047. Rocketdyne SN H-1047
The engine was installed in position 102 on S-I-3 and launched on SA-3 on 16 November 1962.

MSFC SN H-1048. Rocketdyne SN H-1048
The engine was installed in position 103 on S-I-3 and launched on SA-3 on 16 November 1962.

MSFC SN H-1049. Rocketdyne SN H-1049
The engine was installed in position 104 on S-I-3 and launched on SA-3 on 16 November 1962.

MSFC SN H-1050. Rocketdyne SN H-1050
Test Division spare, scrapped.

MSFC SN H-1051. Rocketdyne SN H-1051
The engine was installed in position 105 on S-I-4 and launched on SA-4 on 28 March 1963.

MSFC SN H-1052. Rocketdyne SN H-1052
The engine was installed in position 106 on S-I-4 and launched on SA-4 on 28 March 1963.

MSFC SN H-1053. Rocketdyne SN H-1053
The engine was installed in position 107 on S-I-4 and launched on SA-4 on 28 March 1963.

MSFC SN H-1054. Rocketdyne SN H-1054
The engine was installed in position 101 on S-I-4 and launched on SA-4 on 28 March 1963.

MSFC SN H-1055. Rocketdyne SN H-1055
The engine was installed in position 102 on S-I-4 and launched on SA-4 on 28 March 1963.

MSFC SN H-1056. Rocketdyne SN H-1056
The engine was installed in position 103 on S-I-4 and launched on SA-4 on 28 March 1963.

MSFC SN H-1057. Rocketdyne SN H-1057

The engine was installed in position 104 on S-I-4 and launched on SA-4 on 28 March 1963.

MSFC SN H-1058. Rocketdyne SN H-1058

The engine was installed in position 108 on S-I-4 and launched on SA-4 on 28 March 1963.

MSFC SN H-1059. Rocketdyne SN H-1059

Spare for SA-4. Test Division spare, scrapped.

MSFC SN H-1060. Rocketdyne SN H-1060

The engine is installed on S-I-D at MSFC.

MSFC SN H-1061. Rocketdyne SN H-5001

First production 188,000 pounds thrust engine. Test Division spare, scrapped.

MSFC SN H-1062. Rocketdyne SN H-2001

The engine was installed in position 105 on S-I-5 and launched on SA-5 on 29 January 1964.

MSFC SN H-1063. Rocketdyne SN H-5002

The engine was installed in position 101 on S-I-5 and launched on SA-5 on 29 January 1964.

MSFC SN H-1064. Rocketdyne SN H-5003

The engine was installed in position 102 on S-I-5 and launched on SA-5 on 29 January 1964.

MSFC SN H-1065. Rocketdyne SN H-5004

The engine was installed in position 103 on S-I-5 and launched on SA-5 on 29 January 1964.

MSFC SN H-1066. Rocketdyne SN H-2002

The engine was installed in position 106 on S-I-5 and launched on SA-5 on 29 January 1964.

MSFC SN H-1067. Rocketdyne SN H-2003

The engine was installed in position 107 on S-I-5 and launched on SA-5 on 29 January 1964.

MSFC SN H-1068. Rocketdyne SN H-2004

The engine was installed in position 108 on S-I-5 and launched on SA-5 on 29 January 1964.

MSFC SN H-1069. Rocketdyne SN H-5005

The engine was installed in position 104 on S-I-5, in place of H-5006, and launched on SA-5 on 29 January 1964.

MSFC SN H-1070. Rocketdyne SN H-5006

Originally installed in S-I-5 but replaced by H-5005. Spare for SA-5 and SA-6. On permanent loan to University of Utah.

MSFC SN H-1071. Rocketdyne SN H-2005

The engine was installed in position 105 on S-I-6 and launched on SA-6 on 28 May 1964.

MSFC SN H-1072. Rocketdyne SN H-2006

The engine was installed in position 106 on S-I-6 and launched on SA-6 on 28 May 1964.

MSFC SN H-1073. Rocketdyne SN H-2007

The engine was installed in position 108 on S-I-6, in place of H-2009, and launched on SA-6 on 28 May 1964.

MSFC SN H-1074. Rocketdyne SN H-2008

The engine was installed in position 107 on S-I-6 and launched on SA-6 on 28 May 1964.

MSFC SN H-1075. Rocketdyne SN H-2009

Originally installed in S-I-6 but replaced by H-2007. Spare for SA-5 and SA-6. Transferred to the Smithsonian.

MSFC SN H-1076. Rocketdyne SN H-5007

The engine was installed in position 101 on S-I-6 and launched on SA-6 on 28 May 1964.

MSFC SN H-1077. Rocketdyne SN H-5008

The engine was installed in position 102 on S-I-6 and launched on SA-6 on 28 May 1964.

MSFC SN H-1078. Rocketdyne SN H-5009

The engine was installed in position 103 on S-I-6 and launched on SA-6 on 28 May 1964.

MSFC SN H-1079. Rocketdyne SN H-5010

The engine was installed in position 104 on S-I-6 and launched on SA-6 on 28 May 1964.

MSFC SN H-1080. Rocketdyne SN H-5011

Test Division spare. The engine is installed on S-I-D at MSFC.

MSFC SN H-1081. Rocketdyne SN H-5012

The engine was installed in position 102 on S-I-9 and launched on SA-9 on 16 February 1965.

MSFC SN H-1082. Rocketdyne SN H-2010

The engine was installed in position 105 on S-I-7 and launched on SA-7 on 18 September 1964.

MSFC SN H-1083. Rocketdyne SN H-2011

Originally installed in S-I-7 but replaced by H-2019. Test Division spare. Transferred to the Smithsonian.

MSFC SN H-1084. Rocketdyne SN H-2012

The engine was installed in position 107 on S-I-7 and launched on SA-7 on 18 September 1964.

MSFC SN H-1085. Rocketdyne SN H-2013
Test Division spare, scrapped.

MSFC SN H-1086. Rocketdyne SN H-2014
Test Division spare. Transferred to the Smithsonian.

MSFC SN H-1087. Rocketdyne SN H-5013
The engine was installed in position 101 on S-I-7 and launched on SA-7 on 18 September 1964.

MSFC SN H-1088. Rocketdyne SN H-5014
Originally installed in S-I-7 but replaced by H-5027. Spare. Transferred to R&D, scrapped.

MSFC SN H-1089. Rocketdyne SN H-5015
The engine was installed in position 103 on S-I-7 and launched on SA-7 on 18 September 1964.

MSFC SN H-1090. Rocketdyne SN H-5016
The engine was installed in position 104 on S-I-7 and launched on SA-7 on 18 September 1964.

MSFC SN H-1091. Rocketdyne SN H-5017
Used in R&D then scrapped.

MSFC SN H-1092. Rocketdyne SN H-2015
Originally installed in S-I-7 but replaced by H-2021. Test Division spare. Transferred to the Smithsonian.

MSFC SN H-1093. Rocketdyne SN H-2016
The engine was installed in position 105 on S-I-8 and launched on SA-8 on 25 May 1965.

MSFC SN H-1094. Rocketdyne SN H-2017
Originally installed in S-I-8 but replaced by H-2032. R&D spare. Transferred to the Smithsonian.

MSFC SN H-1095. Rocketdyne SN H-2018
The engine was installed in position 107 on S-I-8 and launched on SA-8 on 25 May 1965.

MSFC SN H-1096. Rocketdyne SN H-2019
The engine was installed in position 106 on S-I-7, in place of H-2011, and launched on SA-7 on 18 September 1964.

MSFC SN H-1097. Rocketdyne SN H-5018
R&D spare. 188,000 pounds thrust PFRT engine, then scrapped.

MSFC SN H-1098. Rocketdyne SN H-5019
The engine was installed in position 101 on S-I-8 and launched on SA-8 on 25 May 1965.

MSFC SN H-1099. Rocketdyne SN H-5020
The engine was installed in position 102 on S-I-8 and launched on SA-8 on 25 May 1965.

MSFC SN H-1100. Rocketdyne SN H-5021
The engine was installed in position 103 on S-I-8 and launched on SA-8 on 25 May 1965.

MSFC SN H-1101. Rocketdyne SN H-5022
Originally installed in S-I-8 but replaced by H-5032. Spare for SA-7 and SA-9. Transferred to the Atlas program.

MSFC SN H-1102. Rocketdyne SN H-2020
The engine was installed in position 105 on S-I-9 and launched on SA-9 on 16 February 1965.

MSFC SN H-1103. Rocketdyne SN H-2021
The engine was installed in position 108 on S-I-7, in place of H-2015, and launched on SA-7 on 18 September 1964.

MSFC SN H-1104. Rocketdyne SN H-2022
The engine was installed in position 106 on S-I-9 and launched on SA-9 on 16 February 1965.

MSFC SN H-1105. Rocketdyne SN H-2023
The engine was installed in position 107 on S-I-9 and launched on SA-9 on 16 February 1965.

MSFC SN H-1106. Rocketdyne SN H-2024
The engine was installed in position 108 on S-I-9 and launched on SA-9 on 16 February 1965.

MSFC SN H-1107. Rocketdyne SN H-5023
The engine was installed in position 101 on S-I-9 and launched on SA-9 on 16 February 1965.

MSFC SN H-1108. Rocketdyne SN H-5024
Spare for SA-8 and SA-10. Transferred to Atlas.

MSFC SN H-1109. Rocketdyne SN H-5025
The engine was installed in position 103 on S-I-9 and launched on SA-9 on 16 February 1965.

MSFC SN H-1110. Rocketdyne SN H-5026
The engine was installed in position 104 on S-I-9 and launched on SA-9 on 16 February 1965.

MSFC SN H-1111. Rocketdyne SN H-5027
The engine was installed in position 102 on S-I-7, in place of H-5014, and launched on SA-7 on 18 September 1964.

MSFC SN H-1112. Rocketdyne SN H-2025
R&D spare. On permanent loan to University of Arkansas.

MSFC SN H-1113. Rocketdyne SN H-2026
The engine was installed in position 106 on S-I-10 and launched on SA-10 on 30 July 1965.

MSFC SN H-1114. Rocketdyne SN H-2027
The engine was installed in position 107 on S-I-10 and launched on SA-10 on 30 July 1965.

MSFC SN H-1115. Rocketdyne SN H-2028
R&D spare. On permanent loan to Missouri School of Mines.

MSFC SN H-1116. Rocketdyne SN H-5028
The engine was installed in position 101 on S-I-10 and launched on SA-10 on 30 July 1965.

MSFC SN H-1117. Rocketdyne SN H-5029
The engine was installed in position 102 on S-I-10 and launched on SA-10 on 30 July 1965.

MSFC SN H-1118. Rocketdyne SN H-5030
The engine was installed in position 103 on S-I-10 and launched on SA-10 on 30 July 1965.

MSFC SN H-1119. Rocketdyne SN H-5031
The engine was installed in position 104 on S-I-10 and launched on SA-10 on 30 July 1965.

MSFC SN H-1120. Rocketdyne SN H-2029
Originally installed in S-I-8 but replaced by H-2031. R&D spare. Transferred to the Smithsonian.

MSFC SN H-1121. Rocketdyne SN H-2030
The engine was installed in position 108 on S-I-10 and launched on SA-10 on 30 July 1965.

MSFC SN H-1122. Rocketdyne SN H-5032
The engine was installed in position 104 on S-I-8, after replacing H-5022, and launched on SA-8 on 25 May 1965.

MSFC SN H-1123. Rocketdyne SN H-2031
The engine was installed in position 108 on S-I-8, after replacing H-2029, and launched on SA-8 on 25 May 1965.

MSFC SN H-1124. Rocketdyne SN H-2032
The engine was installed in position 106 on S-I-8, after replacing H-2017, and launched on SA-8 on 25 May 1965.

MSFC SN H-1125. Rocketdyne SN H-5033
R&D spare, scrapped.

MSFC SN H-1126. Rocketdyne SN H-2033
Spare for SA-8 and SA-10. Used in ageing study. Transferred to Atlas.

MSFC SN H-1127. Rocketdyne SN H-2034
The engine was installed in position 105 on S-I-10 and launched on SA-10 on 30 July 1965.

MSFC SN H-1128. Rocketdyne SN H-2035
Spare, obsolete. Transferred to Atlas.

MSFC SN H-1129. Rocketdyne SN H-5034
Spare, obsolete. On display at MAF.

MSFC SN H-1130. Rocketdyne SN H-5035
Spare, obsolete. Transferred to the Smithsonian.

MSFC SN H-1131. Rocketdyne SN H-5036
The engine is installed on S-I-D at MSFC.

MSFC SN H-1132. Rocketdyne SN H-5037
Spare, obsolete. On display at USSRC.

MSFC SN H-1133. Rocketdyne SN H-2036
Spare for SA-7 and SA-9. Transferred to Atlas.

MSFC SN H-1134. Rocketdyne SN H-2037
Spare, obsolete. R&D spare, scrapped.

MSFC SN H-1135. Rocketdyne SN H-2038
Spare, obsolete. Transferred to the Smithsonian.

MSFC SN H-1136. Rocketdyne SN H-2039
Spare, obsolete. Transferred to Atlas.

MSFC SN H-1137. Rocketdyne SN H-5038
Spare, obsolete. R&D spare, scrapped.

MSFC SN H-1138. Rocketdyne SN H-5039
Spare, obsolete. Transferred to the Smithsonian.

MSFC SN H-1139. Rocketdyne SN H-5040
Spare, obsolete. Transferred to Atlas.

MSFC SN H-1140. Rocketdyne SN H-5041
Spare, obsolete. AS-210 age engine. Transferred to Atlas.

MSFC SN H-1141. Rocketdyne SN H-2040
Spare, obsolete. Transferred to Atlas.

MSFC SN H-1142. Rocketdyne SN H-2041
Spare, obsolete. Transferred to Atlas.

MSFC SN H-1143. Rocketdyne SN H-2042
Spare, obsolete. Transferred to Atlas.

MSFC SN H-1144. Rocketdyne SN H-2043
Spare, obsolete. Transferred to Atlas.

MSFC SN H-1145. Rocketdyne SN H-5042
Spare, R&D. Not assembled. Turned over to R&D for components.

MSFC SN H-1146. Rocketdyne SN H-5043
Spare, R&D. Not assembled. Turned over to R&D for components.

MSFC SN H-1147. Rocketdyne SN H-5044
Spare, R&D. Not assembled. Turned over to R&D for components.

MSFC SN H-1148. Rocketdyne SN H-5045
Spare, R&D. Not assembled. Turned over to R&D for components.

MSFC SN H-1149. Rocketdyne SN H-4044
The engine was accepted by NASA on 28 March 1964. The engine was installed in position 105 on S-IB-1 and launched on AS-201 on 26 February 1966.

MSFC SN H-1150. Rocketdyne SN H-4045
The engine was accepted by NASA on 31 March 1964. The engine was installed in position 106 on S-IB-1 and launched on AS-201 on 26 February 1966.

MSFC SN H-1151. Rocketdyne SN H-4046
The engine was accepted by NASA on 31 March 1964. Originally installed in S-IB-1 but replaced by H-4052. The engine was installed in position 108 on S-IB-2, after replacing H-4051, and launched on AS-202 on 25 August 1966.

MSFC SN H-1152. Rocketdyne SN H-4047
The engine was accepted by NASA on 30 April 1964. The engine was installed in position 108 on S-IB-1 and launched on AS-201 on 26 February 1966.

MSFC SN H-1153. Rocketdyne SN H-7046
The engine was accepted by NASA on 19 June 1964. The engine was installed in position 101 on S-IB-1 and launched on AS-201 on 26 February 1966.

MSFC SN H-1154. Rocketdyne SN H-7047
The engine was accepted by NASA on 30 April 1964. The engine was installed in position 102 on S-IB-1 and launched on AS-201 on 26 February 1966.

MSFC SN H-1155. Rocketdyne SN H-7048
The engine was accepted by NASA on 30 June 1964. The engine was installed in position 103 on S-IB-1 and launched on AS-201 on 26 February 1966.

MSFC SN H-1156. Rocketdyne SN H-7049
The engine was accepted by NASA on 20 June 1964. The engine was installed in position 104 on S-IB-1 and launched on AS-201 on 26 February 1966.

MSFC SN H-1157. Rocketdyne SN H-7050
The engine was accepted by NASA on 3 March 1965. The engine was installed in position 103 on S-IB-2, after replacing H-7053, and launched on AS-202 on 25 August 1966.

MSFC SN H-1158. Rocketdyne SN H-4048
The engine was accepted by NASA on 30 June 1964. The engine was installed in position 105 on S-IB-2 and launched on AS-202 on 25 August 1966.

MSFC SN H-1159. Rocketdyne SN H-4049
The engine was accepted by NASA on 8 April 1964. The engine was installed in position 106 on S-IB-2 and launched on AS-202 on 25 August 1966.

MSFC SN H-1160. Rocketdyne SN H-4050
The engine was accepted by NASA on 31 July 1964. The engine was installed in position 107 on S-IB-2 and launched on AS-202 on 25 August 1966.

MSFC SN H-1161. Rocketdyne SN H-4051
The engine was accepted by NASA on 31 July 1964. Originally installed in S-IB-2 but replaced by H-4046. The engine was a spare for SA-201, SA-202. Scrapped.

MSFC SN H-1162. Rocketdyne SN H-7051
The engine was accepted by NASA on 2 April 1965. The engine was installed in position 101 on S-IB-2 and launched on AS-202 on 25 August 1966.

MSFC SN H-1163. Rocketdyne SN H-7052
The engine was accepted by NASA on 7 April 1965. The engine was installed in position 102 on S-IB-2 and launched on AS-202 on 25 August 1966.

MSFC SN H-1164. Rocketdyne SN H-7053
The engine was accepted by NASA on 31 August 1964. Originally installed in S-IB-2 but replaced by H-7050. FWV spare. Scrapped.

MSFC SN H-1165. Rocketdyne SN H-7054
The engine was accepted by NASA on 2 April 1965. The engine was installed in position 104 on S-IB-2 and launched on AS-202 on 25 August 1966.

MSFC SN H-1166. Rocketdyne SN H-4052
The engine was accepted by NASA on 30 September 1964. The engine was installed in position 107 on S-IB-1, after replacing H-4046, and launched on AS-201 on 26 February 1966.

MSFC SN H-1167. Rocketdyne SN H-4053
The engine was accepted by NASA on 30 January 1965. The engine was installed in position 105 on S-IB-3 and launched on AS-203 on 5 July 1966.

MSFC SN H-1168. Rocketdyne SN H-4054
The engine was accepted by NASA on 30 January 1965. The engine was installed in position 106 on S-IB-3 and launched on AS-203 on 5 July 1966.

MSFC SN H-1169. Rocketdyne SN H-4055

The engine was accepted by NASA on 27 February 1965. Qual engine. Scrapped.

MSFC SN H-1170. Rocketdyne SN H-7055

The engine was accepted by NASA on 27 February 1965. Qual engine. Scrapped.

MSFC SN H-1171. Rocketdyne SN H-4056

The engine was accepted by NASA on 30 January 1965. The engine was installed in position 107 on S-IB-3 and launched on AS-203 on 5 July 1966.

MSFC SN H-1172. Rocketdyne SN H-4057

The engine was accepted by NASA on 13 February 1965. The engine was installed in position 108 on S-IB-3 and launched on AS-203 on 5 July 1966.

MSFC SN H-1173. Rocketdyne SN H-7056

The engine was accepted by NASA on 30 January 1965. The engine was installed in position 101 on S-IB-3 and launched on AS-203 on 5 July 1966.

MSFC SN H-1174. Rocketdyne SN H-7057

The engine was accepted by NASA on 31 October 1964. Test Division spare. Last reported at MSFC.

MSFC SN H-1175. Rocketdyne SN H-7058

The engine was accepted by NASA on 30 January 1965. The engine was installed in position 102 on S-IB-3 and launched on AS-203 on 5 July 1966.

MSFC SN H-1176. Rocketdyne SN H-7059

The engine was accepted by NASA on 30 January 1965. The engine was installed in position 103 on S-IB-3 and launched on AS-203 on 5 July 1966.

MSFC SN H-1177. Rocketdyne SN H-7060

The engine was accepted by NASA on 30 January 1965. The engine was installed in position 104 on S-IB-3 and launched on AS-203 on 5 July 1966.

MSFC SN H-1178. Rocketdyne SN H-7061

The engine was accepted by NASA on 12 February 1965. R&D spare. Scrapped.

MSFC SN H-1179. Rocketdyne SN H-4058

The engine was accepted by NASA on 27 February 1965. The engine was installed in position 105 on S-IB-4 and launched on AS-204 on 22 January 1968.

MSFC SN H-1180. Rocketdyne SN H-4059

The engine was accepted by NASA on 27 February 1965. Originally installed in S-IB-4 but replaced by H-4062. Spare for SA-204, SA-205. Transferred to the RS-27 program.

MSFC SN H-1181. Rocketdyne SN H-4060

The engine was accepted by NASA on 27 February 1965. The engine was installed in position 107 on S-IB-4 and launched on AS-204 on 22 January 1968.

MSFC SN H-1182. Rocketdyne SN H-4061

The engine was accepted by NASA on 27 February 1965. The engine was installed in position 108 on S-IB-4 and launched on AS-204 on 22 January 1968.

MSFC SN H-1183. Rocketdyne SN H-7062

The engine was accepted by NASA on 27 February 1965. The engine was installed in position 101 on S-IB-4 and launched on AS-204 on 22 January 1968.

MSFC SN H-1184. Rocketdyne SN H-7063

The engine was accepted by NASA on 27 February 1965. The engine was installed in position 102 on S-IB-4 and launched on AS-204 on 22 January 1968.

MSFC SN H-1185. Rocketdyne SN H-7064

The engine was accepted by NASA on 28 February 1965. The engine was installed in position 103 on S-IB-4 and launched on AS-204 on 22 January 1968.

MSFC SN H-1186. Rocketdyne SN H-7065

The engine was accepted by NASA on 28 February 1965. The engine was installed in position 104 on S-IB-4 and launched on AS-204 on 22 January 1968.

MSFC SN H-1187. Rocketdyne SN H-4062

The engine was accepted by NASA on 31 March 1965. The engine was installed in position 106 on S-IB-4, after replacing H-4059, and launched on AS-204 on 22 January 1968.

MSFC SN H-1188. Rocketdyne SN H-4063

The engine was accepted by NASA on 25 June 1965. The engine was installed in position 105 on S-IB-5 and launched on AS-205 on 11 October 1968.

MSFC SN H-1189. Rocketdyne SN H-4064

The engine was accepted by NASA on 25 June 1965. The engine was installed in position 106 on S-IB-5 and launched on AS-205 on 11 October 1968.

MSFC SN H-1190. Rocketdyne SN H-4065

The engine was accepted by NASA on 30 June 1965. The engine was installed in position 107 on S-IB-5 and launched on AS-205 on 11 October 1968.

MSFC SN H-1191. Rocketdyne SN H-7066

The engine was accepted by NASA on 30 June 1965. The engine was installed in position 101 on S-IB-5 and launched on AS-205 on 11 October 1968.

MSFC SN H-1192. Rocketdyne SN H-7067

The engine was accepted by NASA on 28 July 1965. The engine was installed in position 102 on S-IB-5 and launched on AS-205 on 11 October 1968.

MSFC SN H-1193. Rocketdyne SN H-4066

The engine was accepted by NASA on 30 July 1965. The engine was installed in position 108 on S-IB-5 and launched on AS-205 on 11 October 1968.

MSFC SN H-1194. Rocketdyne SN H-7068

The engine was accepted by NASA on 30 July 1965. The engine was installed in position 103 on S-IB-5 and launched on AS-205 on 11 October 1968.

MSFC SN H-1195. Rocketdyne SN H-7069

The engine was accepted by NASA on 30 July 1965. The engine was installed in position 104 on S-IB-5 and launched on AS-205 on 11 October 1968.

MSFC SN H-1196. Rocketdyne SN H-4067

The engine was accepted by NASA on 30 August 1965. Test Division spare. Transferred to the Smithsonian.

MSFC SN H-1197. Rocketdyne SN H-7070

The engine was accepted by NASA on 27 August 1965. Spare for AS-204, AS-205. Transferred to the RS-27 program.

MSFC SN H-1198. Rocketdyne SN H-4068

The engine was subject to 3 engine-level acceptance firings at Rocketdyne's Neosho plant, with a cumulative firing time of 240.6 seconds. The final firing took place on 5 October 1965. The engine was accepted by NASA on 21 October 1965. The engine was shipped from Neosho on 26 October 1965, arriving at MAF on 28 October 1965. The engine was installed in position 105 on S-IB-6 on 31 December 1965 and launched on AS-206 on 25 May 1973.

MSFC SN H-1199. Rocketdyne SN H-4069

The engine was subject to 3 engine-level acceptance firings at Rocketdyne's Neosho plant, with a cumulative firing time of 240.7 seconds. The final firing took place on 18 October 1965. The engine was accepted by NASA on 29 October 1965. The engine was shipped from Neosho on 2 November 1965, arriving at MAF on 4 November 1965. The engine was installed in position 106 on S-IB-6 on 7 January 1966 and launched on AS-206 on 25 May 1973.

MSFC SN H-1200. Rocketdyne SN H-7071

The engine was subject to 4 engine-level acceptance firings at Rocketdyne's Neosho plant, with a cumulative firing time of 283.1 seconds. The final firing took place on 13 October 1965. The engine was accepted by NASA on 26 October 1965. The engine was shipped

from Neosho on 26 October 1965, arriving at MAF on 28 October 1965. The engine was installed in position 101 on S-IB-6 on 18 January 1966 and launched on AS-206 on 25 May 1973.

MSFC SN H-1201. Rocketdyne SN H-7072

The engine was subject to 5 engine-level acceptance firings at Rocketdyne's Neosho plant, with a cumulative firing time of 324.8 seconds. The final firing took place on 25 October 1965. The engine was accepted by NASA on 29 October 1965. The engine was shipped from Neosho on 3 November 1965, arriving at MAF on 8 November 1965. The engine was installed in position 102 on S-IB-6 on 25 January 1966 and launched on AS-206 on 25 May 1973.

MSFC SN H-1202. Rocketdyne SN H-4070

The engine was subject to 4 engine-level acceptance firings at Rocketdyne's Neosho plant, with a cumulative firing time of 222.6 seconds. The final firing took place on 21 October 1965. The engine was accepted by NASA on 29 October 1965. The engine was shipped from Neosho on 3 November 1965, arriving at MAF on 8 November 1965. The engine was installed in position 107 on S-IB-6 on 4 January 1966 and launched on AS-206 on 25 May 1973.

MSFC SN H-1203. Rocketdyne SN H-4071

The engine was subject to 7 engine-level acceptance firings at Rocketdyne's Neosho plant, with a cumulative firing time of 498.7 seconds. The final firing took place on 14 July 1966. The engine was initially accepted by NASA on 19 November 1965. The engine was shipped from Neosho on 10 August 1966, arriving at MAF on 12 August 1966. Originally installed in S-IB-6 but replaced by H-4072. The engine was installed in position 106 on S-IB-8 on 23 November 1966, after replacing H-4078, and launched on AS-208 on 16 November 1973.

MSFC SN H-1204. Rocketdyne SN H-7073

The engine was subject to 3 engine-level acceptance firings at Rocketdyne's Neosho plant, with a cumulative firing time of 240.3 seconds. The final firing took place on 8 November 1965. The engine was accepted by NASA on 19 November 1965. The engine was shipped from Neosho on 24 November 1965, arriving at MAF on 29 November 1965. The engine was installed in position 103 on S-IB-6 on 28 January 1966 and launched on AS-206 on 25 May 1973.

MSFC SN H-1205. Rocketdyne SN H-7074

The engine was subject to 5 engine-level acceptance firings at Rocketdyne's Neosho plant, with a cumulative firing time of 362.2 seconds. The final firing took place on 9 August 1966. The engine was initially accepted by NASA on 30 November 1965. The engine was shipped from Neosho on 8 September 1966, arriving at MAF on

12 September 1966. The engine was installed in position 104 on S-IB-7 on 16 November 1966, after replacing H-7080, and launched on AS-207 on 28 July 1973.

MSFC SN H-1206. Rocketdyne SN H-4072

The engine was subject to 3 engine-level acceptance firings at Rocketdyne's Neosho plant, with a cumulative firing time of 240.3 seconds. The final firing took place on 17 February 1966. The engine was accepted by NASA on 28 February 1966. The engine was shipped from Neosho on 11 March 1966, arriving at MAF on 14 March 1966. The engine was installed in position 108 on S-IB-6 on 1 July 1966, after replacing H-4071, and launched on AS-206 on 25 May 1973.

MSFC SN H-1207. Rocketdyne SN H-7075

The engine was subject to 3 engine-level acceptance firings at Rocketdyne's Neosho plant, with a cumulative firing time of 240.0 seconds. The final firing took place on 18 November 1965. The engine was accepted by NASA on 30 November 1965. The engine was shipped from Neosho on 8 December 1965, arriving at MAF on 10 December 1965. The engine was installed in position 104 on S-IB-6 on 26 January 1966 and launched on AS-206 on 25 May 1973.

MSFC SN H-1208. Rocketdyne SN H-7076

The engine was subject to 3 engine-level acceptance firings at Rocketdyne's Neosho plant, with a cumulative firing time of 240.9 seconds. The final firing took place on 10 December 1965. The engine was accepted by NASA on 29 December 1965. The engine was shipped from Neosho on 6 January 1966, arriving at MAF on 10 January 1966. The engine was installed in position 103 on S-IB-7 on 23 July 1966, after replacing H-7079, and launched on AS-207 on 28 July 1973.

MSFC SN H-1209. Rocketdyne SN H-4073

The engine was subject to 11 engine-level acceptance firings at Rocketdyne's Neosho plant, with a cumulative firing time of 707.5 seconds. The final firing took place on 20 January 1967. The engine was initially accepted by NASA on 29 December 1965. The engine was shipped from Neosho on 4 August 1967, arriving at MAF on 7 August 1967. Originally installed in S-IB-7 but replaced by H-4078. The engine was installed in position 105 on S-IB-12 on 28 August 1967, replacing H-4096. Removed from the stage on 13 August 1973 and transferred to the RS-27 program.

MSFC SN H-1210. Rocketdyne SN H-4074

The engine was subject to 3 engine-level acceptance firings at Rocketdyne's Neosho plant, with a cumulative firing time of 240.5 seconds. The final firing took place on 21 December 1965. The engine was accepted by NASA on 29 December 1965. The engine was shipped from Neosho on 11 January 1966, arriving at MAF on 13 January 1966. The engine was installed in

position 106 on S-IB-7 on 28 March 1966 and launched on AS-207 on 28 July 1973.

MSFC SN H-1211. Rocketdyne SN H-7077

The engine was accepted by NASA on 29 December 1965. Originally installed in S-IB-7 but replaced by H-7085. Spare. Disassembled and transferred to the RS-27 program.

MSFC SN H-1212. Rocketdyne SN H-7078

The engine was subject to 7 engine-level acceptance firings at Rocketdyne's Neosho plant, with a cumulative firing time of 575.2 seconds. The final firing took place on 13 January 1967. The engine was initially accepted by NASA on 29 December 1965. The engine was shipped from Neosho on 28 January 1967, arriving at MAF on 30 January 1967. The engine was installed in position 102 on S-IB-7 on 16 February 1967 and launched on AS-207 on 28 July 1973.

MSFC SN H-1213. Rocketdyne SN H-4075

The engine was subject to 3 engine-level acceptance firings at Rocketdyne's Neosho plant, with a cumulative firing time of 240.4 seconds. The final firing took place on 18 January 1966. The engine was accepted by NASA on 31 January 1966. The engine was shipped from Neosho on 2 February 1966, arriving at MAF on 4 February 1966. The engine was installed in position 107 on S-IB-7 on 24 March 1966 and launched on AS-207 on 28 July 1973.

MSFC SN H-1214. Rocketdyne SN H-4076

The engine was subject to 3 engine-level acceptance firings at Rocketdyne's Neosho plant, with a cumulative firing time of 247.1 seconds. The final firing took place on 19 January 1966. The engine was accepted by NASA on 31 January 1966. The engine was shipped from Neosho on 4 February 1966, arriving at MAF on 7 February 1966. The engine was installed in position 108 on S-IB-7 on 23 March 1966 and launched on AS-207 on 28 July 1973.

MSFC SN H-1215. Rocketdyne SN H-7079

The engine was subject to 7 engine-level acceptance firings at Rocketdyne's Neosho plant, with a cumulative firing time of 376.1 seconds. The final firing took place on 24 October 1966. The engine was initially accepted by NASA on 28 January 1966. The engine was shipped from Neosho on 18 November 1966, arriving at MAF on 21 November 1966. Originally installed in S-IB-7 but replaced by H-7076. The engine was installed in position 102 on S-IB-8 on 1 August 1967, after replacing H-7083, and launched on AS-208 on 16 November 1973.

MSFC SN H-1216. Rocketdyne SN H-7080

The engine was accepted by NASA on 31 January

1966. Originally installed in S-IB-7 but replaced by H-7074. FWV spare. Transferred to the RS-27 program.

MSFC SN H-1217. Rocketdyne SN H-7081

The engine was subject to 6 engine-level acceptance firings at Rocketdyne's Neosho plant, with a cumulative firing time of 577.8 seconds. The final firing took place on 9 January 1967. The engine was initially accepted by NASA on 25 February 1966. The engine was shipped from Neosho on 21 January 1967, arriving at MAF on 23 January 1967. The engine was installed in position 103 on S-IB-8 on 2 February 1967, after replacing H-7084, and launched on AS-208 on 16 November 1973.

MSFC SN H-1218. Rocketdyne SN H-4077

The engine was subject to 6 engine-level acceptance firings at Rocketdyne's Neosho plant, with a cumulative firing time of 576.2 seconds. The final firing took place on 29 December 1966. The engine was initially accepted by NASA on 28 February 1966. The engine was shipped from Neosho on 12 January 1967, arriving at MAF on 17 January 1967. The engine was installed in position 105 on S-IB-8 on 26 January 1967 and launched on AS-208 on 16 November 1973.

MSFC SN H-1219. Rocketdyne SN H-4078

The engine was subject to 6 engine-level acceptance firings at Rocketdyne's Neosho plant, with a cumulative firing time of 361.3 seconds. The final firing took place on 2 February 1967. The engine was initially accepted by NASA on 28 February 1966. The engine was shipped from Neosho on 20 March 1967, arriving at MAF on 23 March 1967. Originally installed in S-IB-8 but replaced by H-4071. The engine was installed in position 105 on S-IB-7 on 10 June 1967, after replacing H-4073, and launched on AS-207 on 28 July 1973.

MSFC SN H-1220. Rocketdyne SN H-7082

The engine was subject to 3 engine-level acceptance firings at Rocketdyne's Neosho plant, with a cumulative firing time of 240.9 seconds. The final firing took place on 8 February 1966. The engine was accepted by NASA on 25 February 1966. The engine was shipped from Neosho on 28 February 1966, arriving at MAF on 2 March 1966. The engine was installed in position 101 on S-IB-8 on 20 June 1966 and launched on AS-208 on 16 November 1973.

MSFC SN H-1221. Rocketdyne SN H-7083

The engine was accepted by NASA on 25 February 1966. Originally installed in S-IB-8 but replaced by H-7079. FWV spare. Transferred to the RS-27 program.

MSFC SN H-1222. Rocketdyne SN H-4079

The engine was subject to 4 engine-level acceptance firings at Rocketdyne's Neosho plant, with a cumulative firing time of 243.3 seconds. The final firing took place on 15 March 1966. The engine was accepted by NASA on 31 March 1966. The engine was shipped from Neosho on 1 April 1966, arriving at MAF on 4 April 1966. The engine was installed in position 107 on S-IB-8 on 7 June 1966 and launched on AS-208 on 16 November 1973.

MSFC SN H-1223. Rocketdyne SN H-4080

The engine was subject to 6 engine-level acceptance firings at Rocketdyne's Neosho plant, with a cumulative firing time of 414.4 seconds. The final firing took place on 22 March 1966. The engine was accepted by NASA on 31 March 1966. The engine was shipped from Neosho on 1 April 1966, arriving at MAF on 4 April 1966. The engine was installed in position 108 on S-IB-8 on 8 June 1966 and launched on AS-208 on 16 November 1973.

MSFC SN H-1224. Rocketdyne SN H-7084

The engine was accepted by NASA on 30 March 1966. Originally installed in S-IB-8 but replaced by H-7081. R&D spare. Scrapped.

MSFC SN H-1225. Rocketdyne SN H-7085

The engine was subject to 8 engine-level acceptance firings at Rocketdyne's Neosho plant, with a cumulative firing time of 665.6 seconds. The final firing took place on 5 January 1967. The engine was initially accepted by NASA on 24 March 1966. The engine was shipped from Neosho on 13 January 1967, arriving at MAF on 14 August 1967. Originally installed in S-IB-8 but replaced by H-7096. The engine was installed in position 101 on S-IB-7 on 16 October 1968, after replacing H-7077, and launched on AS-207 on 28 July 1973.

MSFC SN H-1226. Rocketdyne SN H-4081

The engine was subject to 5 engine-level acceptance firings at Rocketdyne's Neosho plant, with a cumulative firing time of 530.1 seconds. The engine was accepted by NASA on 31 March 1966. Installed in position 106 on S-IB-9, after replacing H-4083. Removed on 10 December 1966 and replaced by H-4086. FWV spare for SA-209. Shipped to Canoga Park on 4 April 1973. Sent to KSC 19 February 1974.

MSFC SN H-1227. Rocketdyne SN H-4082

The engine was subject to 3 engine-level acceptance firings at Rocketdyne's Neosho plant, with a cumulative firing time of 241.7 seconds. The final firing took place on 14 April 1966. The engine was accepted by NASA on 26 April 1966. The engine was shipped from Neosho on 27 April 1966, arriving at MAF on 29 April 1966. The engine was installed in position 105 on S-IB-9 on 19 August 1966.

MSFC SN H-1228. Rocketdyne SN H-4083

The engine was accepted by NASA on 30 April 1966. 205,000 pounds thrust re-qual engine. Originally installed in S-IB-9 but replaced by H-4081. Scrapped.

MSFC SN H-1229. Rocketdyne SN H-7086

The engine was subject to 4 engine-level acceptance firings at Rocketdyne's Neosho plant, with a cumulative firing time of 395.5 seconds. The final firing took place on 25 January 1967. The engine was initially accepted by NASA on 22 April 1966. The engine was shipped from Neosho on 10 February 1967, arriving at MAF on 13 February 1967. Originally installed in S-IB-9 but replaced by H-7090. The engine was installed in position 103 on S-IB-11 on 13 June 1967, replacing H-7097.

MSFC SN H-1230. Rocketdyne SN H-7087

The engine was subject to 5 engine-level acceptance firings at Rocketdyne's Neosho plant, with a cumulative firing time of 326.8 seconds. The final firing took place on 19 April 1966. The engine was accepted by NASA on 29 April 1966. The engine was shipped from Neosho on 4 May 1966, arriving at MAF on 5 May 1966. The engine was installed in position 102 on S-IB-9 on 12 September 1966.

MSFC SN H-1231. Rocketdyne SN H-4084

The engine was subject to 3 engine-level acceptance firings at Rocketdyne's Neosho plant, with a cumulative firing time of 241.3 seconds. The final firing took place on 28 April 1966. The engine was accepted by NASA on 13 May 1966. The engine was shipped from Neosho on 3 June 1966, arriving at MAF on 6 June 1966. The engine was installed in position 107 on S-IB-9 on 22 August 1966.

MSFC SN H-1232. Rocketdyne SN H-4085

The engine was subject to 3 engine-level acceptance firings at Rocketdyne's Neosho plant, with a cumulative firing time of 240.8 seconds. The final firing took place on 16 May 1966. The engine was accepted by NASA on 27 May 1966. The engine was shipped from Neosho on 8 June 1966, arriving at MAF on 10 June 1966. The engine was installed in position 108 on S-IB-9 on 22 August 1966.

MSFC SN H-1233. Rocketdyne SN H-7088

The engine was subject to 3 engine-level acceptance firings at Rocketdyne's Neosho plant, with a cumulative firing time of 241.2 seconds. The final firing took place on 13 May 1966. The engine was accepted by NASA on 27 May 1966. The engine was shipped from Neosho on 8 June 1966, arriving at MAF on 10 June 1966. The engine was installed in position 103 on S-IB-9 on 7 September 1966.

MSFC SN H-1234. Rocketdyne SN H-7089

The engine was subject to 5 engine-level acceptance firings at Rocketdyne's Neosho plant, with a cumulative firing time of 289.2 seconds. The final firing took place on 18 May 1966. The engine was accepted by NASA on 27 May 1966. The engine was shipped from Neosho on 3 June 1966, arriving at MAF on 6 June 1966. The engine was installed in position 104 on S-IB-9 on 30 August 1966.

MSFC SN H-1235. Rocketdyne SN H-7090

The engine was subject to 3 engine-level acceptance firings at Rocketdyne's Neosho plant, with a cumulative firing time of 241.3 seconds. The final firing took place on 31 May 1966. The engine was accepted by NASA on 10 June 1966. The engine was shipped from Neosho on 15 June 1966, arriving at MAF on 17 June 1966. The engine was installed in position 101 on S-IB-9 on 5 January 1967 (after replacing H-7086) before itself being replaced by H-7110. Disassembled and transferred to the RS-27 program.

MSFC SN H-1236. Rocketdyne SN H-4086

The engine was subject to 3 engine-level acceptance firings at Rocketdyne's Neosho plant, with a cumulative firing time of 241.0 seconds. The final firing took place on 20 May 1966. The engine was accepted by NASA on 27 May 1966. The engine was shipped from Neosho on 15 June 1966, arriving at MAF on 17 June 1966. The engine was installed in position 106 on S-IB-9 on 30 December 1966, replacing H-4081.

MSFC SN H-1237. Rocketdyne SN H-4087

The engine was subject to 4 engine-level acceptance firings at Rocketdyne's Neosho plant, with a cumulative firing time of 284.2 seconds. The final firing took place on 8 June 1966. The engine was accepted by NASA on 21 June 1966. The engine was shipped from Neosho on 30 June 1966, arriving at MAF on 5 July 1966. The engine was installed in position 105 on S-IB-10 on 27 October 1966 and launched on AS-210 on 15 July 1975.

MSFC SN H-1238. Rocketdyne SN H-4088

The engine was subject to 4 engine-level acceptance firings at Rocketdyne's Neosho plant, with a cumulative firing time of 283.4 seconds. The final firing took place on 16 June 1966. The engine was initially accepted by NASA on 14 June 1966. The engine was shipped from Neosho on 30 June 1966, arriving at MAF on 5 July 1966. The engine was installed in position 106 on S-IB-10 on 25 October 1966 before being replaced by H-4104. Transferred to the RS-27 program.

MSFC SN H-1239. Rocketdyne SN H-4089

The engine was subject to 3 engine-level acceptance firings at Rocketdyne's Neosho plant, with a cumulative firing time of 237.8 seconds. The final firing took place on 17 June 1966. The engine was accepted by NASA on 29 June 1966. The engine was shipped from Neosho on 6 July 1966, arriving at MAF on 8 July 1966. The engine was installed in position 107 on S-IB-10 on 25 October 1966 and launched on AS-210 on 15 July 1975.

MSFC SN H-1240. Rocketdyne SN H-4090

The engine was subject to 3 engine-level acceptance firings at Rocketdyne's Neosho plant, with a cumulative firing time of 240.3 seconds. The final firing took place on 17 June 1966. The engine was accepted by

NASA on 29 June 1966. The engine was shipped from Neosho on 6 July 1966, arriving at MAF on 8 July 1966. The engine was installed in position 108 on S-IB-10 on 27 October 1966 and launched on AS-210 on 15 July 1975.

MSFC SN H-1241. Rocketdyne SN H-7091

The engine was subject to 4 engine-level acceptance firings at Rocketdyne's Neosho plant, with a cumulative firing time of 352.4 seconds. The final firing took place on 1 July 1966. The engine was accepted by NASA on 15 July 1966. The engine was shipped from Neosho on 27 July 1966, arriving at MAF on 29 July 1966. The engine was installed in position 101 on S-IB-10 on 31 October 1966 and launched on AS-210 on 15 July 1975.

MSFC SN H-1242. Rocketdyne SN H-7092

The engine was subject to 5 engine-level acceptance firings at Rocketdyne's Neosho plant, with a cumulative firing time of 443.8 seconds. The final firing took place on 22 February 1967. The engine was initially accepted by NASA on 29 July 1966. The engine was shipped from Neosho on 20 March 1967, arriving at MAF on 23 March 1967. Originally installed in S-IB-10 but replaced by H-7099. The engine was installed in position 101 on S-IB-11 on 8 June 1967, replacing H-7095.

MSFC SN H-1243. Rocketdyne SN H-7093

The engine was subject to 5 engine-level acceptance firings at Rocketdyne's Neosho plant, with a cumulative firing time of 286.0 seconds. The final firing took place on 13 July 1966. The engine was accepted by NASA on 29 July 1966. The engine was shipped from Neosho on 10 August 1966, arriving at MAF on 12 August 1966. The engine was installed in position 103 on S-IB-10 on 3 November 1966 and launched on AS-210 on 15 July 1975.

MSFC SN H-1244. Rocketdyne SN H-7094

The engine was subject to 3 engine-level acceptance firings at Rocketdyne's Neosho plant, with a cumulative firing time of 241.8 seconds. The final firing took place on 18 July 1966. The engine was accepted by NASA on 29 July 1966. The engine was shipped from Neosho on 3 August 1966, arriving at MAF on 5 August 1966. The engine was installed in position 104 on S-IB-10 on 5 December 1966 and launched on AS-210 on 15 July 1975.

MSFC SN H-1245. Rocketdyne SN H-4091

The engine was subject to 6 engine-level acceptance firings at Rocketdyne's Neosho plant, with a cumulative firing time of 463.7 seconds. The final firing took place on 11 July 1966. The engine was accepted by NASA on 26 July 1966. The engine was shipped from Neosho on 27 July 1966, arriving at MAF on 29 July 1966. The engine was installed in position 108 on S-IB-11 on 16 March 1968, replacing H-4095.

MSFC SN H-1246. Rocketdyne SN H-4092

The engine was subject to 4 engine-level acceptance firings at Rocketdyne's Neosho plant, with a cumulative firing time of 286.0 seconds. The final firing took place on 19 September 1966. The engine was initially accepted by NASA on 30 August 1966. The engine was shipped from Neosho on 7 October 1966, arriving at MAF on 10 October 1966. The engine was installed in position 105 on S-IB-11 on 5 January 1967.

MSFC SN H-1247. Rocketdyne SN H-4093

The engine was subject to 3 engine-level acceptance firings at Rocketdyne's Neosho plant, with a cumulative firing time of 241.6 seconds. The final firing took place on 11 August 1966. The engine was accepted by NASA on 31 August 1966. The engine was shipped from Neosho on 8 September 1966, arriving at MAF on 12 September 1966. The engine was installed in position 106 on S-IB-11 on 13 January 1967.

MSFC SN H-1248. Rocketdyne SN H-7095

The engine was subject to 3 engine-level acceptance firings at Rocketdyne's Neosho plant, with a cumulative firing time of 243.4 seconds. The final firing took place on 28 July 1966. The engine was accepted by NASA on 24 August 1966. The engine was shipped from Neosho on 2 June 1967, arriving at MAF on 5 June 1967. Originally installed in S-IB-11 but replaced by H-7092. The engine was installed in position 102 on S-IB-11 on 14 July 1967, replacing H-7096.

MSFC SN H-1249. Rocketdyne SN H-7096

The engine was subject to 3 engine-level acceptance firings at Rocketdyne's Neosho plant, with a cumulative firing time of 243.2 seconds. The final firing took place on 2 August 1966. The engine was accepted by NASA on 24 August 1966. The engine was shipped from Neosho on 25 August 1966, arriving at MAF on 19 June 1967. Originally installed in S-IB-11 but replaced by H-7095. The engine was installed in position 104 on S-IB-8 on 10 July 1967, after replacing H-7085, and launched on AS-208 on 16 November 1973.

MSFC SN H-1250. Rocketdyne SN H-4094

The engine was subject to 3 engine-level acceptance firings at Rocketdyne's Neosho plant, with a cumulative firing time of 243.5 seconds. The final firing took place on 16 August 1966. The engine was accepted by NASA on 2 September 1966. The engine was shipped from Neosho on 15 September 1966, arriving at MAF on 19 September 1966. The engine was installed in position 107 on S-IB-11 on 17 January 1967.

MSFC SN H-1251. Rocketdyne SN H-4095

The engine was accepted by NASA on 7 September 1966. Originally installed in S-IB-11 but replaced by H-4091. R&D engine, used for re-build of HT6B engine.

MSFC SN H-1252. Rocketdyne SN H-7097

The engine was subject to 4 engine-level acceptance firings at Rocketdyne's Neosho plant, with a cumulative firing time of 269.7 seconds. The engine was accepted by NASA on 9 September 1966. Received at MAF on 19 September 1966. Engine installed in 103 position in S-IB-11 stage on 17 February 1967. Removed from the stage at MAF on 12 May 1967 and shipped to Neosho on 31 May 1967. Replaced by H-7086. Returned to MAF 9 August 1967. FWV engine for S-IB-12 25 June 1969. Shipped to MSFC 15 July 1970. Shipped to MAF 1 October 1971. FWV engine for AS-210 12 June 1973. Transferred to CCSD 22 January 1974. Shipped to KSC 19 April 1974. Shipped to Canoga Park 13 November 1974. Shipped to EFL 13 December 1974. Last reported at EFL.

MSFC SN H-1253. Rocketdyne SN H-7098

The engine was subject to 3 engine-level acceptance firings at Rocketdyne's Neosho plant, with a cumulative firing time of 243.5 seconds. The final firing took place on 24 August 1966. The engine was accepted by NASA on 15 September 1966. The engine was shipped from Neosho on 21 July 1967, arriving at MAF on 24 July 1967. Originally installed in S-IB-11 but replaced by H-7102. The engine was installed in position 103 on S-IB-12 on 22 September 1967, replacing H-7102. Removed from the stage on 2 August 1973 and transferred to the RS-27 program.

MSFC SN H-1254. Rocketdyne SN H-7099

The engine was subject to 3 engine-level acceptance firings at Rocketdyne's Neosho plant, with a cumulative firing time of 243.4 seconds. The final firing took place on 26 August 1966. The engine was accepted by NASA on 15 September 1966. The engine was shipped from Neosho on 7 October 1966, arriving at MAF on 10 October 1966. The engine was installed in position 102 on S-IB-10 on 20 December 1966, after replacing H-7092, and launched on AS-210 on 15 July 1975.

MSFC SN H-1255. Rocketdyne SN H-4096

The engine was accepted by NASA on 13 October 1966. Originally installed in S-IB-12 but replaced by H-4073. Spare. Disassembled and transferred to the RS-27 program.

MSFC SN H-1256. Rocketdyne SN H-4097

The engine was subject to 3 engine-level acceptance firings at Rocketdyne's Neosho plant, with a cumulative firing time of 220.6 seconds. The final firing took place on 3 October 1966. The engine was accepted by NASA on 17 October 1966. The engine was shipped from Neosho on 10 November 1966, arriving at MAF on 14 November 1966. The engine was installed in position 106 on S-IB-12 on 15 March 1967. Removed from the stage on 10 August 1973 and transferred to the RS-27 program.

MSFC SN H-1257. Rocketdyne SN H-7100

The engine was subject to 3 engine-level acceptance firings at Rocketdyne's Neosho plant, with a cumula-

tive firing time of 282.5 seconds. The final firing took place on 28 September 1966. The engine was accepted by NASA on 17 October 1966. The engine was shipped from Neosho on 21 October 1966, arriving at MAF on 24 October 1966. The engine was installed in position 101 on S-IB-12 on 11 September 1967. Removed from the stage on 3 August 1973 and transferred to the RS-27 program.

MSFC SN H-1258. Rocketdyne SN H-7101

The engine was subject to 3 engine-level acceptance firings at Rocketdyne's Neosho plant, with a cumulative firing time of 241.4 seconds. The final firing took place on 5 October 1966. The engine was accepted by NASA on 18 October 1966. The engine was shipped from Neosho on 10 November 1966, arriving at MAF on 14 November 1966. The engine was installed in position 102 on S-IB-12 on 8 September 1967. Removed from the stage on 1 August 1973 and transferred to the RS-27 program.

MSFC SN H-1259. Rocketdyne SN H-4098

The engine was subject to 2 engine-level acceptance firings at Rocketdyne's Neosho plant, with a cumulative firing time of 200.6 seconds. The final firing took place on 1 November 1966. The engine was accepted by NASA on 15 November 1966. The engine was shipped from Neosho on 21 December 1966, arriving at MAF on 23 December 1966. The engine was installed in position 107 on S-IB-12 on 16 March 1967. Removed from the stage on 1 August 1973 and transferred to the RS-27 program.

MSFC SN H-1260. Rocketdyne SN H-4099

The engine was subject to 4 engine-level acceptance firings at Rocketdyne's Neosho plant, with a cumulative firing time of 365.7 seconds. The final firing took place on 5 December 1966. The engine was accepted by NASA on 20 December 1966. The engine was shipped from Neosho on 21 December 1966, arriving at MAF on 23 December 1966. The engine was installed in position 108 on S-IB-12 on 10 March 1967. Removed from the stage on 2 August 1973 and transferred to the RS-27 program.

MSFC SN H-1261. Rocketdyne SN H-7102

The engine was subject to 2 engine-level acceptance firings at Rocketdyne's Neosho plant, with a cumulative firing time of 200.7 seconds. The final firing took place on 27 October 1966. The engine was accepted by NASA on 15 November 1966. The engine was shipped from Neosho on 18 November 1966, arriving at MAF on 21 November 1966. Originally allocated to S-IB-12 but replaced by H-7098. The engine was installed in position 104 on S-IB-11 on 8 June 1967, replacing H-7098.

MSFC SN H-1262. Rocketdyne SN H-7103

The engine was subject to 3 engine-level acceptance firings at Rocketdyne's Neosho plant, with a cumulative firing time of 282.9 seconds. The final firing took place on 13 December 1966. The engine was accepted by NASA on 29 December 1966. The engine was shipped from Neosho on 10 February 1967, arriving at MAF on 13 February 1967. The engine was installed in position 104 on S-IB-12 on 28 September 1967. Removed from the stage on 3 August 1973 and transferred to the RS-27 program.

MSFC SN H-1263. Rocketdyne SN H-4100
The engine was accepted by NASA on 29 June 1967. Disassembled and transferred to the RS-27 program.

MSFC SN H-1264. Rocketdyne SN H-4101
The engine was accepted by NASA on 29 June 1967. Disassembled and transferred to the RS-27 program.

MSFC SN H-1265. Rocketdyne SN H-7104
The engine was accepted by NASA on 29 June 1967. Disassembled and transferred to the RS-27 program.

MSFC SN H-1266. Rocketdyne SN H-7105
The engine was accepted by NASA on 29 June 1967. Disassembled and transferred to the RS-27 program.

MSFC SN H-1267. Rocketdyne SN H-4102
The engine was accepted by NASA on 29 June 1967. Disassembled and transferred to the RS-27 program.

MSFC SN H-1268. Rocketdyne SN H-4103
The engine was accepted by NASA on 29 June 1967. Disassembled and transferred to the RS-27 program.

MSFC SN H-1269. Rocketdyne SN H-7106
The engine was accepted by NASA on 29 June 1967. Disassembled and transferred to the RS-27 program.

MSFC SN H-1270. Rocketdyne SN H-7107
The engine was accepted by NASA on 29 June 1967. Disassembled and transferred to the RS-27 program.

MSFC SN H-1271. Rocketdyne SN H-4104
The engine was subject to 3 engine-level acceptance firings at Rocketdyne's Neosho plant, with a cumulative firing time of 215.5 seconds. The final firing took place on 8 May 1967. The engine was accepted by NASA on 29 June 1967. The engine was shipped from Neosho on 29 June 1967, arriving at MAF on 3 July 1967. The engine was installed in position 106 on S-IB-10 on 16 February 1973, after replacing H-4088, and launched on AS-210 on 15 July 1975.

MSFC SN H-1272. Rocketdyne SN H-4105
The engine was subject to 3 engine-level acceptance firings at Rocketdyne's Neosho plant, with a cumulative firing time of 282.5 seconds. The final firing took place on 23 May 1967. The engine was accepted by NASA on 29 June 1967. The engine was shipped from Neosho on 10 July 1967, arriving at MAF on 13 July 1967. The engine was installed in position 105 on S-IB-14 on 14 April 1969. Removed from the stage on 9 May 1973

and transferred to the RS-27 program.

MSFC SN H-1273. Rocketdyne SN H-4106
The engine was accepted by NASA on 29 June 1967. Disassembled and transferred to the RS-27 program.

MSFC SN H-1274. Rocketdyne SN H-7108
The engine was accepted by NASA on 29 June 1967. Disassembled and transferred to the RS-27 program.

MSFC SN H-1275. Rocketdyne SN H-7109
The engine was accepted by NASA on 31 July 1967. Disassembled and transferred to the RS-27 program.

MSFC SN H-1276. Rocketdyne SN H-4107
The engine was subject to 3 engine-level acceptance firings at Rocketdyne's Neosho plant, with a cumulative firing time of 239.0 seconds. The engine was accepted by NASA on 27 July 1967. FWV engine for S-IB-13 stage on 23 June 1969. Shipped to MSFC 17 July 1970. Shipped to MAF 8 February 1972. FWV engine for S-IB-11 on 27 June 1973. Last reported at MAF.

MSFC SN H-1277. Rocketdyne SN H-4108
The engine was subject to 2 engine-level acceptance firings at Rocketdyne's Neosho plant, with a cumulative firing time of 197.9 seconds. The engine was accepted by NASA on 28 July 1967. Flight spare for AS-210. Transferred to CCSD on 1 September 1971. Last reported at MAF.

MSFC SN H-1278. Rocketdyne SN H-7110
The engine was subject to 3 engine-level acceptance firings at Rocketdyne's Neosho plant, with a cumulative firing time of 239.4 seconds. The final firing took place on 11 July 1967. The engine was accepted by NASA on 18 August 1967. The engine was shipped from Neosho on 25 August 1967, arriving at MAF on 28 August 1967. The engine was installed in position 102 on S-IB-9 on 27 October 1972, after replacing H-7090.

MSFC SN H-1279. Rocketdyne SN H-7111
The engine was subject to 6 engine-level acceptance firings at Rocketdyne's Neosho plant, with a cumulative firing time of 493.1 seconds. The final firing took place on 16 August 1967. The engine was accepted by NASA on 31 August 1967. The engine was shipped from Neosho on 18 September 1967, arriving at MAF on 20 September 1967. The engine was installed in position 101 on S-IB-14 on 18 April 1969. Removed from the stage on 30 April 1973 and transferred to the RS-27 program.

MSFC SN H-1280. Rocketdyne SN H-7112
The engine was subject to 2 engine-level acceptance firings at Rocketdyne's Neosho plant, with a cumulative firing time of 238.0 seconds. The engine was accepted by NASA on 28 August 1967. Flight spare for SA-209, AS-210, SA-211. Transferred to CCSD 1 September 1971. Last reported at MAF.

MSFC SN H-1281. Rocketdyne SN H-4109

The engine was subject to 3 engine-level acceptance firings at Rocketdyne's Neosho plant, with a cumulative firing time of 280.1 seconds. The final firing took place on 12 September 1967. The engine was accepted by NASA on 29 September 1967. The engine was shipped from Neosho on 13 October 1967, arriving at MAF on 16 October 1967. The engine was installed in position 105 on S-IB-13 on 5 March 1969. Removed from the stage on 26 January 1973 and became a flight spare for AS-210. Shipped to KSC on 19 April 1974. Last reported at KSC.

MSFC SN H-1282. Rocketdyne SN H-4110

The engine was subject to 3 engine-level acceptance firings at Rocketdyne's Neosho plant, with a cumulative firing time of 292.4 seconds. The final firing took place on 14 September 1967. The engine was accepted by NASA on 29 September 1967. The engine was shipped from Neosho on 13 October 1967, arriving at MAF on 16 October 1967. The engine was installed in position 106 on S-IB-13 on 6 March 1969. Removed from the stage on 7 February 1973 and transferred to the RS-27 program.

MSFC SN H-1283. Rocketdyne SN H-7113

The engine was subject to 4 engine-level acceptance firings at Rocketdyne's Neosho plant, with a cumulative firing time of 320.6 seconds. The final firing took place on 24 August 1967. The engine was accepted by NASA on 15 September 1967. The engine was shipped from Neosho on 18 September 1967, arriving at MAF on 20 September 1967. The engine was installed in position 101 on S-IB-13 on 25 March 1969. Removed from the stage on 8 February 1973 and became a flight spare for AS-210 on 14 February 1973. Shipped to KSC 19 April 1974. Last reported at KSC.

MSFC SN H-1284. Rocketdyne SN H-7114

The engine was subject to 3 engine-level acceptance firings at Rocketdyne's Neosho plant, with a cumulative firing time of 254.2 seconds. The final firing took place on 26 September 1967. The engine was accepted by NASA on 13 October 1967. The engine was shipped from Neosho on 20 October 1967, arriving at MAF on 23 October 1967. The engine was installed in position 102 on S-IB-13 on 14 March 1969. Removed from the stage on 30 January 1973 and transferred to the RS-27 program.

MSFC SN H-1285. Rocketdyne SN H-4111

The engine was subject to 3 engine-level acceptance firings at Rocketdyne's Neosho plant, with a cumulative firing time of 216.1 seconds. The final firing took place on 9 October 1967. The engine was accepted by NASA on 24 October 1967. The engine was shipped from Neosho on 16 November 1967, arriving at MAF on 20 November 1967. The engine was installed in position 107 on S-IB-13 on 6 March 1969. Removed from the stage on 2 February 1973 and transferred to the RS-27 program.

MSFC SN H-1286. Rocketdyne SN H-4112

The engine was subject to 3 engine-level acceptance firings at Rocketdyne's Neosho plant, with a cumulative firing time of 197.6 seconds. The final firing took place on 16 October 1967. The engine was accepted by NASA on 9 November 1967. The engine was shipped from Neosho on 16 November 1967, arriving at MAF on 20 November 1967. The engine was installed in position 108 on S-IB-13 on 7 March 1969. Removed from the stage on 5 February 1973 and transferred to the RS-27 program.

MSFC SN H-1287. Rocketdyne SN H-7115

The engine was subject to 4 engine-level acceptance firings at Rocketdyne's Neosho plant, with a cumulative firing time of 396.3 seconds. The final firing took place on 3 October 1967. The engine was accepted by NASA on 19 October 1967. The engine was shipped from Neosho on 20 October 1967, arriving at MAF on 23 October 1967. The engine was installed in position 103 on S-IB-13 on 13 March 1969. Removed from the stage on 6 February 1973 and transferred to the RS-27 program.

MSFC SN H-1288. Rocketdyne SN H-7116

The engine was subject to 4 engine-level acceptance firings at Rocketdyne's Neosho plant, with a cumulative firing time of 246.0 seconds. The final firing took place on 13 November 1967. The engine was accepted by NASA on 29 November 1967. The engine was shipped from Neosho on 1 December 1967, arriving at MAF on 4 December 1967. The engine was installed in position 104 on S-IB-13 on 24 March 1969. Removed from the stage on 29 January 1973 and transferred to the RS-27 program.

MSFC SN H-1289. Rocketdyne SN H-4113

The engine was subject to 2 engine-level acceptance firings at Rocketdyne's Neosho plant, with a cumulative firing time of 198.6 seconds. The final firing took place on 20 November 1967. The engine was accepted by NASA on 13 December 1967. The engine was shipped from Neosho on 26 December 1967, arriving at MAF on 28 December 1967. The engine was installed in position 106 on S-IB-14 on 17 April 1969. Removed from the stage on 10 May 1973 and transferred to the RS-27 program.

MSFC SN H-1290. Rocketdyne SN H-4114

The engine was subject to 2 engine-level acceptance firings at Rocketdyne's Neosho plant, with a cumulative firing time of 198.2 seconds. The final firing took place on 1 December 1967. The engine was accepted by NASA on 20 December 1967. The engine was shipped from Neosho on 26 December 1967, arriving at MAF on 28 December 1967. The engine was installed in position 107 on S-IB-14 on 16 April 1969. Removed from the stage on 8 May 1973 and transferred to the RS-27 program.

MSFC SN H-1291. Rocketdyne SN H-7117

The engine was subject to 2 engine-level acceptance firings at Rocketdyne's Neosho plant, with a cumulative firing time of 197.9 seconds. The final firing took place on 31 October 1967. The engine was accepted by NASA on 28 November 1967. The engine was shipped from Neosho on 1 December 1967, arriving at MAF on 4 December 1967. The engine was installed in position 102 on S-IB-14 on 24 April 1969. Removed from the stage on 1 May 1973 and transferred to the RS-27 program.

MSFC SN H-1292. Rocketdyne SN H-7118

The engine was subject to 4 engine-level acceptance firings at Rocketdyne's Neosho plant, with a cumulative firing time of 281.6 seconds. The final firing took place on 15 January 1968. The engine was accepted by NASA on 30 January 1968. The engine was shipped from Neosho on 2 February 1968, arriving at MAF on 5 February 1968. The engine was installed in position 103 on S-IB-14 on 22 April 1969. Removed from the stage on 2 May 1973 and transferred to the RS-27 program.

MSFC SN H-1293. Rocketdyne SN H-4115

The engine was subject to 2 engine-level acceptance firings at Rocketdyne's Neosho plant, with a cumulative firing time of 198.1 seconds. The final firing took place on 9 December 1967. The engine was accepted by NASA on 26 December 1967. The engine was shipped from Neosho on 30 January 1968, arriving at MAF on 1 February 1968. The engine was installed in position 108 on S-IB-14 on 17 April 1969. Removed from the stage on 9 May 1973 and transferred to the RS-27 program.

MSFC SN H-1294. Rocketdyne SN H-7119

The engine was subject to 3 engine-level acceptance firings at Rocketdyne's Neosho plant, with a cumulative firing time of 272.3 seconds. The final firing took place on 11 January 1968. The engine was accepted by NASA on 30 January 1968. The engine was shipped from Neosho on 2 February 1968, arriving at MAF on 5 February 1968. The engine was installed in position 104 on S-IB-14 on 23 April 1969. Removed from the stage on 3 May 1973 and transferred to the RS-27 program.

J-2 production line at Rocketdyne's Canoga Park (1964)

J-2 ENGINE

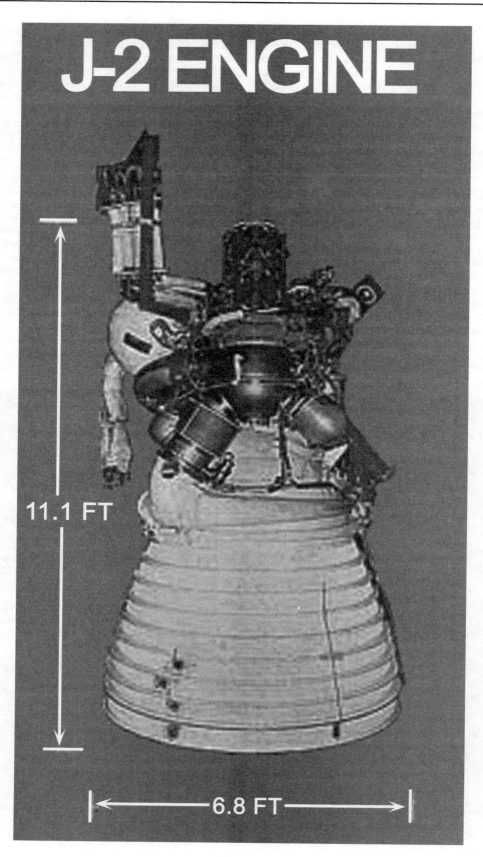

11.1 FT

|← 6.8 FT →|

J-2 Engine

Introduction

The J-2 engine was developed to provide the thrust for the Saturn IB second stage, the Saturn V second stage and the Saturn V third stage. Five J-2 engines were used in the Saturn V second stage and one engine in each of the other applications. In the Saturn V configuration a restart capability was required in order to propel the Apollo spacecraft on Trans-lunar Injection (TLI). Each engine had a thrust of 200,000 lbs, later upgraded to 230,000 lbs. Fuel was liquid hydrogen and the oxidizer was liquid oxygen. The engine was designed by Rocketdyne at its Canoga Park facility in northwestern Los Angeles.

J-2 production line, Canoga Park (9.3.1966)

Background development history

The R&D contract for the J-2 engine, NAS8-19, was awarded to Rocketdyne in 1960. The first test of an un-cooled thrust chamber took place at Santa Susana Field Laboratory on 11 November 1960. The basic engine design was completed on 1 March 1961, at which time the VTS-1 Test Stand at SSFL also was completed and the first test of the gas generator took place.

The first test of the fuel turbo-pump took place on 9 November 1961 and the first test of the oxidizer turbo-pump was on 14 November 1961. The VTS-3B Test Stand was completed on 22 November 1961. The first test of the gas generator system was on 24 January 1962 with a first duration test of this system on 1 March 1962. The VTS-3A Test Stand was completed on 26 March 1962. The first Block I engine static firing test took place at the VTS-3B Test Stand on 31 January 1962. Also during 1962 VTS-2 Test Stand and Delta-2A and -2B Test Stands were completed.

A contract for production of 55 flight engines, NAS8-5603, was awarded on 24 June 1964. On 24 August 1964 NASA announced that it would purchase another 102 J-2 engines. Development engine firings took place at Rocketdyne's test facility at Santa Susana Field Laboratory (SSFL), not far from their design and manufacturing headquarters at Canoga Park in Los Angeles.

Design improvements during the development phase included the addition of a rigid T-ring device to reinforce the thrust chamber to prevent distortion that occurred during the stresses of repeated firings. During May 1964 engine systems tests concentrated on elimination of fuel pump stall at flight stage inlet conditions. The major component problem to resolve in this time period concerned performance of the fuel turbo-pump. Various thrust chamber injectors were tested in order to obtain a stable injector for the FRT engine.

Other engine components tested and improved at SSFL included the LOX pump inducer, fuel pump diaphragm, augmented spark igniter, and heat exchanger.

Flight standard engines

It was planned that Rocketdyne initiate PFRT testing in June 1964 using engine J-2004, or J-2005 as a back up. However both engines were damaged during acceptance testing and had to be returned to Rocketdyne.

PFRT testing was accomplished during November 1964 using engine J-2008. Testing was performed on Test Stands Delta 2A and 2B at SSFL and comprised 16 tests for 2,350 seconds of main-stage duration. On 11 December 1964 a major milestone in the J-2 program was achieved when a ground test engine demonstrated restart capability. After a firing of 165 seconds, and a simulated coast of 75 minutes the engine was restarted for 7 seconds, followed by a further six-minute shutdown and a final restart firing of 310 seconds. FRT testing was performed with engines, J-2023 and J-2022 in mid-1965. Qualification 1 was completed on 13 December 1965 using engine J-2032. The engine was fired a total of 30 times with a cumulative duration of 3,774 seconds. On 14 March 1966 Rocketdyne successfully completed the reliability demonstration program for the 230,000 lb configuration of the J-2 engine. Qualification 2 was completed on 22 August 1966 using engine J-2072.

In total Rocketdyne manufactured and delivered 152 production J-2 engines. A breakdown of the engine disposition is as follows.

65 engines were launched on S-II stages
9 engines were launched on S-IVB-200 stages
12 engines were launched on S-IVB-500 stages
10 engines are on S-II flight stages not launched
2 engines are on S-IVB-500 stages not launched

J-2 production line, Canoga Park (28.1.1964)

1 engine is on an S-IVB-200 stage not launched
37 engines are either in storage or on display in a
museum or unaccounted
5 engines are on the S-II-F/D museum stage
11 engines were scrapped or consumed or stripped for
spare parts

Engine serialization

J-2 engines were serialized sequentially from J-2001 to
J-2152, with no distinction being made between
engines for Saturn IB or Saturn V applications.

Engine configuration changes

There were six major J-2 configurations. Summarized
below are the main changes between configurations;

Configuration
Rocketdyne s/no.

1
Pre-production R&D

2
J-2001 to J-2011

3
J-2012 to J-2019

4
J-2020 to J-2059

5
J-2060 to J-2139

6
J-2140 to J-2152

Configuration 1

Early R&D engine design was for a tank head start.
Stall problems and subsequent overheating necessitated a change to an engine mounted start tank.
Helium control gases had to be kept at a low temperature to prevent venting off of gas. A helium tank integral with the hydrogen start tank eliminated this problem.
An oxidizer turbine bypass valve was added to better
regulate the rate of increase in oxidizer pump speed.

Configuration 2

These engines were ground test engines.

Configuration 3

First production-type engines.
Bleed system was increased to 1.5" diameter.
Gas generator was made integral with the fuel pump
and incorporated an improved film coolant combustor.
The spark de-energize timer was reduced from 5 sec
to 3 sec and the start tank discharge delay timer
changed from 0.50 to 0.64 sec.
Added accumulator to the pneumatic control system.

J-2 ready for hot fire at Santa Susana (1960s)

Configuration 4

Insulated start tank.
Incorporated high performance T/C injector.

J-2 hot fire test at SSFL (1960s)

Electrical control assembly gold plated circuit boards replaced with solder plated boards.
Thrust chamber painted white – deleted requirement at J-2038 and subsequent, retrofit stripping of paint from fuel inlet manifold to exit end of chamber.
Added stage static instrumentation.
Added redundant main stage OK pressure switch.
Incorporated armored harness.

Configuration 5

Engine thrust up-rated to 230,000 lbs.
Incorporated thermostatic control orifice in closing side of main oxidizer valve actuator, rotated sequence valve.
Up-rated oxidizer and fuel turbo pumps.

Configuration 6

Eliminated excessive and redundant instrumentation.

Engine usage

The first R&D engines were not serialized and were used to test the J-2 concept during tests at SSFL. Engines J-2001 to J-2014 were used for ground testing at engine level and in ground test stages. Engine J-2015 was the first engine to be launched, on S-IVB-201 on 22 February 1966. All other engines were used on a combination of S-II, S-IVB-200 and S-IVB-500 stages as well as for ground test and as spares.

Engine logistics

All J-2 engines were manufactured and assembled at Rocketdyne's design and manufacturing plant at Canoga Park. Each engine was trucked to Rocketdyne's nearby test center at Santa Susana Field Laboratory (SSFL) for acceptance testing. SSFL included a test stand for simulated altitude testing by means of a steam

J-2 hot fire test at SSFL (19.11.1963)

ejector system. The engines returned to Canoga Park by truck for final delivery readiness checks and evaluation of test results. Following acceptance by NASA each engine was transported to its next destination that was generally Seal Beach for S-II stage integration or Huntington Beach for S-IVB stage integration. Both of these locations are just a few hours drive from Canoga Park in greater Los Angeles. Some engines were installed in the Battleship stages that were tested at SSFL, MSFC, SACTO and AEDC.

Listing of all J-2 engines associated with the Saturn IB program

Engine J-2001

The engine was manufactured and assembled by Rocketdyne at Canoga Park. It was the first flight standard J-2 engine manufactured. It was subjected to acceptance firings at the Santa Susana Field Laboratory (SSFL). The engine was accepted on 26 November 1963. It became an R&D spare in the Production Support Program. Accountability was transferred on 21 February 1967 to the Smithsonian museum and the engine was later installed, for display purposes, in the S-II-F/D stage that has been on display at the Space and Rocket Center, Huntsville, since 1969.

Engine J-2003

The engine was manufactured and assembled by Rocketdyne at Canoga Park. It was subjected to 15 acceptance firings at the Santa Susana Field Laboratory (SSFL) and started production hot fire acceptance tests with a firing on 13 March 1964. The engine was accepted on 24 April 1964 and shipped to SACTO, Sacramento for use in the S-IVB/IB Battleship on 30 April 1964. It was installed in the stage on 4 June 1964. It was present for 4 Battleship firings between 1 and 23 December 1964. It was removed from the battleship on 28 January 1965 and was transferred to "R&D spare" classification on 28 October 1965. The engine was last reported to be in Canoga Park. By the completion of all testing this engine had accumulated 1526 seconds in 21 firings.

Engine J-2006

The engine was manufactured and assembled by Rocketdyne at Canoga Park. It was rebuilt from J-201 simulator parts. It was acceptance tested with 3 firings at the Santa Susana Field Laboratory (SSFL) totaling 291 seconds and accepted by NASA on 12 June 1964. It was initially delivered to NAA at Seal Beach for S-II battleship use but was later diverted and delivered to DAC at Huntington Beach and installed on the S-IVB-D stage. Later it was used for R&D spares before being transferred to the Smithsonian museum on 6 January 1969. Throughout its life this engine had a cumulative firing time of 291 seconds in 3 firings, all at engine level.

Engine J-2008

The engine was manufactured and assembled by Rocketdyne at Canoga Park.
It was acceptance tested at the Santa Susana Field Laboratory (SSFL) and accepted by NASA on 9 October 1964. On 17 November 1965 the engine was transferred to the R&D group at Rocketdyne and was last reported to be in Canoga Park. It has been used for Preliminary Flight Rating Testing. Throughout its life this engine had a cumulative firing time of 2355 seconds in 18 firings.

Engine J-2012

The engine was manufactured and assembled by Rocketdyne at Canoga Park.
It was acceptance tested with 3 firings at the Santa Susana Field Laboratory (SSFL), totaling 381 seconds, and accepted by NASA on 2 November 1964. It was briefly installed in the S-IVB-201 flight stage in order to checkout systems prior to the arrival of the flight engine. It is the baseline production engine and has been used as an R&D spare. On 14 May 1969 the

engine was transferred to the ASRC in Huntsville where it was on a single engine display. It was subsequently installed in the S-II-F/D stage on display at the Davidson Center at USSRC. Throughout its life this engine had a cumulative firing time of 381 seconds in 3 firings.

Engine J-2013

The engine was manufactured and assembled by Rocketdyne at Canoga Park. It was subjected to acceptance firings at the Santa Susana Field Laboratory (SSFL). The engine was accepted on 20 January 1965 and shipped to SACTO, Sacramento for use in the S-IVB/IB Battleship. It was installed in the battleship on 28 January 1965. It was present for 7 Battleship firings between 13 March and 4 May 1965. It was then transferred to MSFC for use in the MSFC S-IVB Battleship. It was present for 6 firings between 2 August and 15 September 1965. It was then transferred to MSFC as a test spare. On 5 June 1969 ownership was transferred to the Smithsonian museum and the engine was installed, for display purposes, in the S-II-D stage that has been on display at the Space and Rocket Center, Huntsville, since 1969. By the completion of all testing this engine had accumulated 2988 seconds in 32 firings.

Engine J-2015

The engine was manufactured and assembled by Rocketdyne at Canoga Park. It was subjected to 7 engine–level acceptance firings with a total firing time of 548 seconds. The engine was accepted on 16 March 1965 and delivered to DAC at Huntington Beach. It was installed in the S-IVB-201 stage and eventually launched on AS-201 flight on 26 February 1966.

Engine J-2016

The engine was manufactured and assembled by Rocketdyne at Canoga Park. It was subjected to 9 engine–level acceptance firings with a total firing time of 509.8 seconds. The engine was accepted on 23 March 1965 and delivered to DAC at Huntington Beach. It was installed in the S-IVB-202 stage and eventually launched on AS-202 flight on 25 August 1966.

Engine J-2019

The engine was manufactured and assembled by Rocketdyne at Canoga Park. It was subjected to 6 engine–level acceptance firings with a total firing time of 637.1 seconds. The engine was accepted on 6 April 1965 and delivered to DAC at Huntington Beach. It was installed in the S-IVB-203 stage and eventually launched on AS-203 flight on 5 July 1966.

Engine J-2020

The engine was manufactured and assembled by Rocketdyne at Canoga Park. It was subject to 3 acceptance firings at SSFL and accepted by NASA on 22 April 1965. It was shipped to SACTO, Sacramento for use in the S-IVB/V Battleship. It was installed in the battleship in May/June 1965 and was present for 10 battleship firings between 19 June and 20 August 1965. It was then removed from the S-IVB battleship in Sacramento, and transferred to the S-II battleship located at SSFL in northern Los Angeles. It was installed in the S-II Battleship stage in the center 205 position in December 1965 at SSFL. It remained in place for the series of five-engine firings between December 1965 and March 1966. Later the engine was transferred to the KSC museum where it is currently on show as a single engine display. Throughout its life this engine had a cumulative firing time of 3674 seconds in 26 firings, 3 of which were at engine level, 10 in the S-IVB/V Battleship and 13 in the S-II Battleship.

Engine J-2022

The engine was manufactured and assembled by Rocketdyne at Canoga Park. It was acceptance tested at the Santa Susana Field Laboratory (SSFL), with a first firing on 4 May 1965, and accepted by NASA on 30 June 1965. It is the first J-2 to be rated at 225,000 lbs thrust. The engine was used for Flight Rating Testing (FRT 2) before being disassembled for engineering and supply support. Accountability was transferred on 15 December 1965. Throughout its life this engine had a cumulative firing time of 5526 seconds in 94 firings.

Engine J-2023

The engine was manufactured and assembled by Rocketdyne at Canoga Park.
It was acceptance tested at the Santa Susana Field Laboratory (SSFL) and accepted by NASA on 19 June 1965. The engine was used for Flight Rating Testing (FRT 1) before being disassembled for engineering and supply support. Accountability was transferred on 13 February 1969. Throughout its life this engine had a cumulative firing time of 2753 seconds in 25 firings.

Engine J-2025

The engine was manufactured and assembled by Rocketdyne at Canoga Park. It was subjected to 5 engine–level acceptance firings with a total firing time of 498.3 seconds. The last firing was on 14 June 1965. The engine was accepted on 24 June 1965 and delivered to DAC at Huntington Beach on 16 July 1965. It was installed in the S-IVB-204 stage on 30 September 1965 and eventually launched on AS-204 flight on 22 January 1968.

Engine J-2027

The engine was manufactured and assembled by Rocketdyne at Canoga Park. It was subjected to acceptance firings at the Santa Susana Field Laboratory (SSFL). The engine was accepted on 24 June 1965 and shipped to MSFC for use in the MSFC S-IVB Battleship. It was present for a number of Battleship firings from 29 October 1965. It was then allocated to MSFC as a test engine on 13 February 1969. By the completion of all testing this engine had accumulated 3578 seconds in 15 firings.

Engine J-2032

The engine was manufactured and assembled by Rocketdyne at Canoga Park.
It was acceptance tested at the Santa Susana Field Laboratory (SSFL) and accepted by NASA on 31 August 1965. The engine was used for Qualification Testing (Qual 1), which was completed on 13 December 1965. It was then disassembled for engineering and supply support. Accountability was transferred on 13 February 1969. Throughout its life this engine had a cumulative firing time of 3775 seconds in 31 firings.

Engine J-2033

The engine was manufactured and assembled by Rocketdyne at Canoga Park. It was subjected to 5 engine–level acceptance firings with a total firing time of 448.9 seconds. The last firing was on 28 July 1965. The engine was accepted on 14 August 1965 and delivered to DAC at Huntington Beach on 17 August 1965. It was installed in the S-IVB-205 stage on 11 January 1966 and eventually launched on AS-205 flight on 11 October 1968.

Engine J-2037

The engine was manufactured and assembled by Rocketdyne at Canoga Park beginning in the second quarter of 1964. It was acceptance tested with 8 firings, totaling 575.3 seconds, at the Santa Susana Field Laboratory (SSFL). The final firing was on 1 September 1965 and the engine was accepted by NASA on 25 September 1965. The engine was originally allocated to the Qualification Test Program. On 18 September 1968 components were modified and the engine was assigned to Storage Life Evaluation Engineering. It was shipped to MSFC on 20 March 1969. On 19 May 1971 it was transferred to Huntington Beach, and then shipped to Seal Beach for storage on 9 July 1971. It was shipped to Canoga Park on 11 July 1972, and accountability was transferred on 4 August 1972. Throughout its life this engine had a cumulative firing time of 1122 seconds in 10 firings.

Engine J-2046

The engine was manufactured and assembled by Rocketdyne at Canoga Park. It was subjected to 5 engine–level acceptance firings with a total firing time of 439.7 seconds. The last firing was on 5 November 1965. The engine was accepted on 20 December 1965 and delivered to DAC at Huntington Beach on 21 December 1965. It was installed in the S-IVB-206 stage on 31 March 1966 and eventually launched on AS-206 flight on 25 May 1973.

Engine J-2048

The engine was manufactured and assembled by Rocketdyne at Canoga Park. It was subjected to acceptance firings at the Santa Susana Field Laboratory (SSFL). The engine was accepted on 17 December 1965 and shipped to MSFC for use in the MSFC S-IVB Battleship. It was transferred to MSFC as a test engine on 15 May 1969. By the completion of all testing this engine had accumulated 3636 seconds in 25 firings.

Engine J-2049

The engine was manufactured and assembled by Rocketdyne at Canoga Park beginning in the fourth quarter of 1964. It was acceptance tested at the Santa Susana Field Laboratory (SSFL) and accepted by NASA on 27 December 1965. Two days later the engine was shipped to Huntington Beach. It was returned to Canoga Park on 4 October 1966 for rework of the oxidizer T/P. Following rework and further acceptance firings the engine had accumulated 520.4 seconds in 6 firings. The last firing was on 9 March 1967. The engine was then shipped to MTF on 10 July 1967. It was returned to Canoga Park on 16 May 1969. It then went to EFL for storage on 6 April 1970. Finally, it was shipped to Canoga Park for strip down on 24 February 1975 to assess age effects in support of the ASTP. This was because ASTP was shortly to launch the S-IVB-210 stage which had never been tested at stage level and the engine, J-2087, had not been fired since 20 September 1966, nine years previously. Throughout its life this engine had a cumulative firing time of 520.4 seconds in 6 firings.

Engine J-2050

The engine was manufactured and assembled by Rocketdyne at Canoga Park.
It was acceptance tested at the Santa Susana Field Laboratory (SSFL) and accepted by NASA on 14 February 1966. The engine was used as an R&D spare and on the MSFC S-IVB Battleship before being transferred to KSC on 21 May 1969 where it is on single engine display. Throughout its life this engine had a cumulative firing time of 9444 seconds in 35 firings.

Engine J-2054

The engine was manufactured and assembled by Rocketdyne at Canoga Park beginning in the fourth quarter of 1964. It was acceptance tested at the Santa Susana Field Laboratory (SSFL), with a cumulative firing time of 655.4 seconds in 3 firings. The final firing was on 17 January 1966 and the engine was accepted by NASA on 15 February 1966. The engine left Canoga Park on 17 February 1966. It was shipped to EFL for storage on 10 September 1970. It was allocated as FWV support to S-IVB-211, but that stage never flew, and engine ownership was transferred to the Smithsonian. The engine is currently on display at the Rocketdyne LLC at Canoga Park. Throughout its life this engine had a cumulative firing time of 655.4 seconds in 3 firings.

Engine J-2056

The engine was manufactured and assembled by Rocketdyne at Canoga Park. It was subjected to 3 engine–level acceptance firings with a total firing time of 418.5 seconds. The last firing was on 27 January 1966. The engine was accepted on 22 February 1966 and delivered to DAC at Huntington Beach on 25 February 1966. It was installed in the S-IVB-207 stage on 11 May 1966 and eventually launched on AS-207 flight on 28 July 1973.

Engine J-2060

The engine was manufactured and assembled by Rocketdyne at Canoga Park beginning in the fourth quarter of 1964. It was acceptance tested at the Santa Susana Field Laboratory (SSFL), with a cumulative firing time of 389.3 seconds in 3 firings. The first firing was on 2 March 1966 and the final firing was on 4 March 1966. This was the first engine to be acceptance tested at the thrust level of 230,000 lbs. The engine was accepted by NASA on 29 March 1966 and left Canoga Park on 8 April 1966. The engine was last reported to be at MSFC. Throughout its life this engine had a cumulative firing time of 5628.4 seconds in 54 firings.

Engine J-2062

The engine was manufactured and assembled by Rocketdyne at Canoga Park. It was subjected to 4 engine–level acceptance firings with a total firing time of 520.2 seconds. The last firing was on 24 March 1966. The engine was accepted on 26 April 1966 and delivered to DAC at Huntington Beach on 27 April 1966. It was installed in the S-IVB-208 stage on 10 August 1966 and eventually launched on AS-208 flight on 16 November 1973.

Engine J-2065

The engine was manufactured and assembled by Rocketdyne at Canoga Park. It was acceptance tested with 3 firings at SSFL and accepted by NASA on 29 April 1966. It was installed in the S-II Battleship stage in the center 205 position in February 1967 at SSFL. It remained in place for the series of five-engine firings between February 1967 and September 1968. On 15 May 1969 the engine was transferred to the Production Support Program at Rocketdyne, Canoga Park. It was assigned as support to the ASTP program, and eventually, with MSFC ownership, now resides at the Stennis Space Center visitor Center (Stennisphere). Throughout its life this engine had a cumulative firing time of 3375 seconds in 21 firings, 3 of which were at engine level and 18 in the S-II Battleship.

Engine J-2072

The engine was manufactured and assembled by Rocketdyne at Canoga Park.
It was acceptance tested at the Santa Susana Field Laboratory (SSFL) and accepted by NASA on 14 July 1966. The engine was used for Qualification Testing (Qual 2), which was completed on 22 August 1966. Accountability was transferred on 13 February 1969. The engine was last reported to be at Canoga Park. Throughout its life this engine had a cumulative firing time of 3807 seconds in 30 firings.

Engine J-2073

The engine was manufactured and assembled by Rocketdyne at Canoga Park.
It was acceptance tested at the Santa Susana Field Laboratory (SSFL) and accepted by NASA on 28 July 1966. The engine was used as an R&D spare. Accountability was transferred on 13 February 1969. The engine was last reported to be at Canoga Park. Throughout its life this engine had a cumulative firing time of 3940 seconds in 30 firings.

Engine J-2083

The engine was manufactured and assembled by Rocketdyne at Canoga Park. It was subjected to 4 engine–level acceptance firings with a total firing time of 432.0 seconds. The last firing was on 26 August 1966. The engine was accepted on 7 October 1966 and delivered to DAC at Huntington Beach on 20 October 1966. It was installed in the S-IVB-209 stage on 15 November 1966. This stage is on display at KSC.

Engine J-2087

The engine was manufactured and assembled by Rocketdyne at Canoga Park. It was subjected to 4 engine–level acceptance firings with a total firing time of 456.9 seconds. The last firing was on 20 September 1966. The engine was accepted on 28 October 1966 and delivered to DAC at Huntington Beach on 11 November 1966. It was installed in the S-IVB-210 stage on 20 January 1967 and eventually launched on AS-210 flight on 15 July 1975.

Engine J-2095

The engine was manufactured and assembled by Rocketdyne at Canoga Park. It was subjected to 3 engine–level acceptance firings with a total firing time of 355.4 seconds. The last firing was on 9 November 1966. The engine was accepted on 16 December 1966 and delivered to DAC at Huntington Beach on 3 January 1967. It was installed in the S-IVB-211 on 11 April 1967. The stage is on display at the USSRC in Huntsville. However, the engine was removed and is on display at the Herman Oberth museum in Germany.

Engine J-2103

The engine was manufactured and assembled by Rocketdyne at Canoga Park starting in the second quarter of 1966. It was subjected to 3 acceptance firings totaling 615.4 seconds at the Santa Susana Field Laboratory (SSFL). The final firing was on 25 January 1967. The engine was accepted on 22 February 1967 and shipped by truck to Huntington Beach on 23 February 1967. The engine was installed in the S-IVB-212 stage. In April 1969 the engine was removed from the stage after it was decided to use this stage as the vehicle for the Skylab space station. The engine was transferred to FWV engineering in support of S-IVB-210 and S-IVB-211 at Huntington Beach. Engine checkout was completed on 2 September 1971 at Huntington Beach. The engine was shipped to EFL for storage on 27 January 1972. It was then shipped to KSC on 25 March 1974 for storage. It was placed in storage with the S-IVB-513 stage on 5 June 1974. It was taken out of storage on 23 October 1974 and checkout was completed on 2 December 1974. It was reported that the engine was stripped of parts under the auspices of the XRS-2200 program. More recently, in 2007 and 2008, the engine power pack supported J-2X testing at Stennis Space Center. This engine accumulated 615.4 seconds in 3 single engine firings.

Engine J-2111

The engine was manufactured and assembled by Rocketdyne at Canoga Park beginning in the second quarter of 1966. It was acceptance tested at the Santa Susana Field Laboratory (SSFL), with a cumulative firing time of

614.8 seconds in 3 firings. The final firing was on 5 April 1967 and the engine was accepted by NASA on 26 May 1967. The engine left Canoga Park on 28 July 1967. It was shipped from EFL to Huntington Beach on 22 June 1971. Engine checkout was completed on 26 July 1971 and the engine was shipped to Seal Beach on 7 September 1971. It was placed in environmental storage with the S-IVB-513 stage on 8 September 1971. It was removed from storage on 12 November 1971 and again placed in long term storage with the S-IVB-513 stage on 4 October 1972. It was removed from storage on 7 June 1973 and shipped to Canoga Park on 14 June 1973 for up-rated static firings in support of ASTP. It was shipped to EFL for storage on 16 January 1974. Finally it was shipped to KSC for storage on 8 August 1974, as an ASTP spare, and eventually was stripped of components for the XRS-2200 program. Throughout its life this engine had a cumulative firing time of 1282.1 seconds in 11 firings.

Engine J-2120

The engine was manufactured and assembled by Rocketdyne at Canoga Park, starting in the fourth quarter of 1966. It was acceptance tested with 5 firings at SSFL with a cumulative firing time of 1046.7 seconds. The final firing was on 27 July 1967. The engine was accepted by NASA on 13 September 1967. It was installed in the S-II Battleship stage in the 202 position in June 1968. It remained in place for the series of five-engine firings between June 1968 and September 1968. It was assigned as support to the ASTP program. On 4 January 1971 its accountability was transferred to Rocketdyne at Canoga Park. Eventually, with MSFC ownership, it was sent to Tinker Air Force Base. Throughout its life this engine had a cumulative firing time of 2072 seconds in 13 firings, 5 of which were at engine level and 8 in the S-II Battleship.

Engine J-2131

The engine was manufactured and assembled by Rocketdyne at Canoga Park. It was acceptance tested at the Santa Susana Field Laboratory (SSFL), with a cumulative firing time of 346 seconds in 3 firings, and accepted by NASA on 19 March 1968. The engine was used by the Quality Assurance department and disassembled for spare parts. The hardware was restocked. Engine accountability was transferred on 15 May 1969. Throughout its life this engine had a cumulative firing time of 346 seconds in 3 firings.

Engine J-2133

The engine was manufactured and assembled by Rocketdyne at Canoga Park beginning in the first quarter of 1967. It was acceptance tested at the Santa Susana Field Laboratory (SSFL), with a cumulative firing time of

496.9 seconds in 4 firings. The final firing was on 13 March 1968 and the engine was accepted by NASA on 21 May 1968. The engine left Canoga Park on 5 June 1968. It was shipped from EFL to Huntington Beach on 15 July 1971. Engine checkout was completed on 12 August 1971. It was placed in environmental storage with the S-IVB-514 stage on 8 September 1971. It was removed from storage on 20 December 1971. It was placed in KSC long term storage on 3 October 1972 with S-IVB-211. It was removed from storage on 7 June 1973 and shipped to Canoga Park on 14 June 1973 for up-rated hot firing in support of ASTP. It was shipped to EFL on 16 January 1974 for storage, and on to KSC for storage on 26 September 1974. It was kept there as a spare for ASTP but later stripped of components for the XRS-2200 program. Throughout its life this engine had a cumulative firing time of 1166.9 seconds in 9 firings.

Engine J-2146

The engine was manufactured and assembled by Rocketdyne at Canoga Park beginning in the fourth quarter of 1967. It was acceptance tested at the Santa Susana Field Laboratory (SSFL), with a cumulative firing time of 348.6 seconds in 3 firings. The final firing was on 24 June 1969 and the engine was accepted by NASA on 29 July 1969. The engine left Canoga Park on 31 July 1969 bound for Huntington Beach. It was shipped to EFL for storage on 5 January 1972. It was shipped to KSC on 31 October 1973. Engine checkout was completed on 20 September 1974, and it was placed in storage there 3 days later as a spare for S-IVB-200 stages. Eventually it was stripped of components for the XRS-2200 program. Throughout its life this engine had a cumulative firing time of 348.6 seconds in 3 firings.

Engines for S-IB-13 and S-IB-14 and spares being stored at Michoud. (10.1968) 6870815

RL10 Engine

The RL10 was the world's first liquid hydrogen/liquid oxygen rocket engine. The engine, designed by Pratt & Whitney, was an evolution of the 304 turbine engine which had been developed briefly between 1956 and 1958, and the J57, an existing hydrocarbon-powered engine that had been modified to burn hydrogen.

In order to develop the 304 engine Pratt & Whitney had established a research facility in southern Florida. ARPA and the Air Force concluded that the Florida facilities and hydrogen experience of Pratt & Whitney should result in the company being awarded a contract for the development of a 15,000 lb LOX/LH2 rocket engine which was initially designated the LR115.

Design of the Pratt & Whitney RL10 engine started in October 1958. The first experimental thrust chamber was tested at rated conditions in May 1959. The first engine firing occurred in July the same year.

Following an early burn-through problem a redesign occurred and the modified engine was tested successfully in September 1959.

Most of the RL10 engines for the S-IV program were manufactured and tested at the Florida Research and Development Center (FRDC) of Pratt & Whitney at West Palm Beach. Development firings were also performed at the NASA Lewis Research Center where the original injector designs were formulated and where variable thrust firings were explored. MSFC also had a single engine RL10 test stand and conducted numerous firings.

The Pre-Flight Rating Test, performed on engine FX-121-9, was completed in April 1960 and comprised 2,442 seconds and 20 starts. On 10 August 1960 MSFC contracted with Pratt & Whitney for the LR119 rocket engine which would be used in a 17,500 lb 4-engine configuration on the S-IV stage.

The S-IV configuration was later changed to 6 x 15,000 lb LR115 engines which became known as the RL10. The RL10 was fired successfully 230 times before being tested in a 2-engine Centaur Battleship configuration on the E5 test stand.

The first dual engine Centaur configuration firing took place on the E5 stand on 6 November 1960. An attempted firing on the following day resulted in an explosion that destroyed the engines. The rebuilt configuration was ready for test in January 1961 but the first firing attempt also resulted in an explosion. The ensuing investigation proved that LOX was introduced into the thrust chamber in such a way that it bypassed the igniter. In the vertical orientation the LOX fell to the bottom of the diffuser,

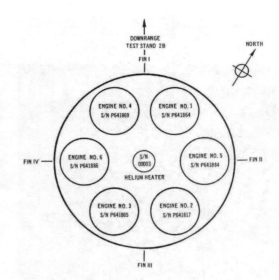

RL10 engine location in S-IV stages

mixed with hydrogen vapor and exploded.

Following modifications to the ignition system, 103 horizontal ignition tests and 50 vertical firings a successful full duration firing of the pair of Centaur RL10 engines on the E5 stand took place on 24 April 1961.

Prior to formal Preliminary Flight Rating Tests (PFRT) testing Pratt and Whitney performed some 712 RL10 development firings accumulating 70,000 seconds of firing time. On 4 November 1961 Pratt and Whitney

RL10 production line at the Florida Research and Development Center at West Palm Beach (Early 1960s)

RL10A-3 engine (2006)

Pratt and Whitney RL10 engine (Early 1960s)

completed Preliminary Flight Rating Tests of the 15,000 pounds thrust RL10A-1 engine, with testing comprising 26 firings with a cumulative firing time of 3,204 seconds. Testing lasted 5 days.

The RL10 was initially utilized in a double-engine configuration on the Centaur upper stage. The first Centaur test vehicle was static fired in March 1962. However, the first Centaur launch, on 8 May 1962, ended in failure after 54 seconds of flight, when the Centaur's hydrogen tank ruptured before the RL10 engines fired.

Between 4 and 9 June 1962 Pratt and Whitney completed Preliminary Flight Rating Tests of the RL10A-3 engine, with testing comprising 26 firings with a cumulative duration of 3,204 seconds. The RL10A-3 engine, used on the S-IV stage, operated at higher pressures than the RL10A-1.

As the production rate increased manufacturing was augmented at the Pratt & Whitney Aircraft East Hartford, Connecticut, plant. The first RL10A-3 engines to be produced at this plant were completed in July 1962.

Full-scale engine throttleability was demonstrated in September 1962. The first successful operation in space, on the Centaur, was on 27 November 1963.

The first flight of the RL10A-3 in the six-engine S-IV stage configuration took place on 29 January 1964, just days after the explosive destruction of the All Systems Vehicle at SACTO.

The Advanced Engine (RL10A-3-1) passed PFRT in September 1964. The RL10 has continued in production to the present time and is base-lined for use on the Altair (Lunar Surface Access Module) lunar lander.

RL10 assembly and test, West Palm Beach (1960s) 9808561

Pregnant Guppy

The Pregnant Guppy was the first in a family of cargo transport aircraft designed specifically to carry large aerospace vehicles. The Pregnant Guppy was manufactured from a modified Boeing Stratocruiser. The maiden flight of the third Stratocruiser prototype, B-377 NX1024V took place on 7 October 1948.

The aircraft was used by Pan American and variously called Clipper Bald Eagle and Clipper Cathay. It was sold back to Boeing in 1961 and eventually to Aero Spacelines, whose Jack Conroy had the idea of enlarging the aircraft and using it for cargo transportation. The Pregnant Guppy first flew in its new configuration on 19 September 1962. Further extensions took place following a viewing by NASA Marshall Management.

In May 1963 MSFC awarded Aero Spacelines, Inc., a contract covering use of the Pregnant Guppy during the period 28 May 1963 through 31 July 1963. The Pregnant Guppy visited Huntsville when it landed at Redstone Airfield in June 1963. On 10 July 1963 the FAA formally certified the Pregnant Guppy aircraft. During July and August MSFC finished its evaluation of the B-377 aircraft for S-IV transportation. Aero Spacelines Inc., owner of the B-377 completed 100 hours of flight test with the craft to determine its suitability for transporting the S-IV stage. The test flights included one cross-country round trip flight between 25 July and 2 August 1963 during part of which the Pregnant Guppy carried the DAC-instrumented S-IV Dynamics/Facilities vehicle.

During August 1963 additional take-off and landing tests were performed which indicated that the Santa Monica airport could be used in place of Los Angeles International Airport, thereby saving half a day's road transport.

On 6 September 1963 MSFC awarded Aero Spacelines a contract for air transportation services through 30 June 1964. The Pregnant Guppy was first used to transport a flight stage when it carried the S-IV-5 stage from Sacramento to Cape Canaveral on 20 September 1963. It was used for the following 2 years to ferry all the remaining S-IV stages. Operating certification was received from the FAA on 13 November 1963. In addition the aircraft was used for the transport of F-1 engines as well as S-IV stages.

The Pregnant Guppy was sold to American Jet Industries in 1974 and continued to carry cargo until it was scrapped in 1979.

Pregnant Guppy at Mather Air Force Base (20.9.1963)

Super Guppy

Following the success of the S-IV flights in the Pregnant Guppy NASA needed an aircraft with a wider diameter to carry the larger S-IVB stage. NASA contracted Aero Spacelines to construct the 377SG Super Guppy, initially dubbed the Very Pregnant Guppy.

It was constructed from Clipper Constitution N1038V, which had been delivered to Pan American on 29 September 1949. Renamed Clipper Hotspur it was sold to Boeing in 1960 and on to Aero Spacelines. Unlike the Pregnant Guppy, the Super Guppy was modified in one stage. The Super Guppy first flew, from Van Nuys airport in Los Angeles, on 31 August 1965.

On 25 September 1965 during a cross country test flight the nose section failed resulting in some damage. NASA awarded an initial contract with Aero Spacelines for use of the Super Guppy on 22 December 1965. The first training flight was on 3 January 1966 when the Super Guppy flew the S-IVB dummy stage from Mojave, California to Los Alamitos Naval Air Station in Los Angeles.

The Super Guppy was first used to transport a flight-type stage on 20 March 1966 when the S-IVB-D stage was transported from MSFC to Los Alamitos Naval Air Station in Los Angeles. The Super Guppy was used to ferry most of the S-IVB stages around the country and continued to be used as a carrier for many years after. The aircraft is currently in storage at Pima Air Force Base in Arizona.

Super Guppy (1966)

Evening take off from Mather Air Force Base (12.1968) 6972000

AKD Point Barrow

Built at the Maryland Shipbuilding and Drydock Company, Baltimore, MD, the Point Barrow was launched on 25 May 1957. She was placed in service by the Military Sea Transportation Service (MSTS) for Arctic service, as Point Barrow (T-AKD-1) on 28 May 1958.

In March 1963 MSFC arranged with the Military Sea Transport Service (MSTS) for use of the AKD Point Barrow in transportation trials. The trials were successfully completed on 15 March 1963. Design work on the modifications needed to the ship took place from January to March 1964. Modifications to the Point Barrow to allow the vessel to undertake Ocean journeys with the S-IVB stages were completed in May 1965 and it was first used to transport a stage the following month.

The Point Barrow was used primarily to transport Saturn stages and engines between Seal Beach Naval docks, Michoud Assembly Facility and Kennedy Space Center, although it did make other trips in support of the Saturn program. The Point Barrow provided a stable, covered transport facility for the vital rocket hardware.

Around 1974 it was reclassified as a Deep Submergence Support Ship and renamed Point Loma (T-AGDS-2). She was struck from the Naval Register on 28 September 1993 and later scrapped.

The S-IB-5 stage on board the Point Barrow leaving Michoud for KSC (25.3.1968) 6864856

Barge Palaemon

The covered barge Palaemon arrived at MSFC for the first time on 22 November 1960, having travelled from its place of construction, Houston, Texas, where it left 16 days previously. The SA-T1 stage was transported to the barge Palaemon for a "shakedown cruise" practice trip with the new barge. The practice trip was conducted to acquaint the crew with the handling of the barge and to permit studies of the stresses to which the rocket would be subjected during its journey to Cape Canaveral. The Palaemon left the MSFC dock on 14 March on its first training trip down the Tennessee River, carrying the SA-T1 stage. After a three-day, 225 mile trip along the Tennessee River, the barge Palaemon returned to the Redstone Arsenal Dock (MSFC) on 17 March.

The barge embarked on its first trial run to the Cape, leaving Huntsville on 17 April 1961. It carried as cargo the S-V dummy stage for the SA-1 flight and ballast simulating the S-I-1 stage. However, it was hit on the port side by a Norwegian tanker on 21 April and the damage necessitated that the Palaemon had to enter a New Orleans shipyard for repairs. These were completed on 25 April and the barge continued on to the Cape. After arriving successfully on 1 May and off-loading the S-V dummy, the Palaemon departed two days later for the return journey to Huntsville, where it arrived at 0600 CDT on 15 May.

Together with the Promise the Palaemon was used for transporting S-1 and S-1B stages, primarily between MSFC and MAF, but occasionally also to KSC. The Palaemon became ice-bound near Paducah, Kentucky, for three weeks in January 1963.

Between 1 August and 2 November 1964 the Palaemon was modified for use transporting the S-IB stages. The modifications included installation of a pilot house and wing bridge. The Palaemon was used for transporting Saturn stages up until 1968.

The Palaemon leaving Michoud for MSFC carrying the S-IB-11 stage (20.10.1967) 6760495

Barge Promise

Following the collapse of the Wheeler Dam on 2 June 1961 MSFC obtained a Navy barge which had been mothballed at Pensacola, Florida. Emergency modifications were made to the barge so that it could carry the SA-1 stages. The barge was renamed Compromise. Later, on 14 December 1961, MSFC completed modifications to the barge, including provision of a cover for the cargo, and renamed it Promise. Promise first transported a Saturn stage on 17 February 1962.

Together with the Palaemon the Promise was used for transporting S-1 and S-1B stages, primarily between MSFC and MAF, but occasionally also to KSC. In September 1965 NASA requested Navy assistance in salvaging the Promise when it became beached on the Michoud levee as a result of Hurricane Betsy on 9 September. The Promise was used for transporting Saturn stages up until 1967.

Barge Orion

Early in January 1965 the Harbor Boat Building Company completed modification of the West Coast Barge (YFNB-20), Orion for use in shipping S-IVB stages along the California coast. Modifications included providing cover for the cargo. From February 1965 the Orion was used for transporting S-IVB stages between Seal Beach and Courtland.

Later Orion was mostly used to ferry Saturn V first and second stages from Michoud and MTF to KSC. It is still used today to transport External Tanks from Michoud to KSC.

The Compromise with S-I stage, dummy S-IV and dummy payload on board leaving Wheeler Dam (5.8.1961)

The Promise after conversion from the Compromise (Mid 1960s)

The first visit of the Orion to MSFC (9.1967) 6758869

Background

In April 1957 members of the Army Ballistic Missile Agency (ABMA) initiated studies to establish possible vehicle configurations to launch a payload of 20,000 to 40,000 pounds for orbital missions and 6,000 to 12,000 pounds for escape missions.

By July 1958, representatives of the Advanced Research Projects Agency (ARPA) showed interest in a clustered booster that would achieve 1,500,000 pounds thrust with available engine hardware. ARPA formally initiated the development program by issuing ARPA Order 14-59 on 15 August 1958. The immediate goal was to demonstrate the feasibility of the engine clustering concept with a full scale captive test firing, using Rocketdyne H-1 engines and available propellant containers. In September 1958, ARPA extended the program to include four test flights of the booster. ARPA Order 47-59, dated 11 December 1958, requested that the Army Ordnance Missile Command (AOMC) design, construct and modify the ABMA captive test tower and associated facilities for booster development, and determine design criteria for suitable launch facilities.

In November 1958, ARPA approved the development of a clustered booster to serve as the first stage of the multi-stage carrier vehicle capable of performing advanced space missions. The project was unofficially known as Juno V until, on 3 February 1959, an ARPA memorandum made the name "Saturn" official.

As the result of the presidential order proposing transfer of the Development Operations Division of ABMA to the National Aeronautics and Space Administration (NASA), an interim agreement was reached between ARPA, NASA and the Department of Defense on 25 November 1959. The agreement provided for the transfer of technical direction of the Saturn program to NASA and for the retention of administrative direction by ARPA. ABMA transferred Saturn program responsibility to NASA on 1 July 1960.

On 28 July 1960 The Douglas Aircraft Company was awarded a contract to develop and fabricate the second stage (S-IV) of the recommended configuration. The original design concept specified four Pratt & Whitney 17,500 pounds thrust liquid hydrogen/liquid oxygen engines (LR119). The design was later modified to utilize six Pratt & Whitney 15,000 pounds thrust liquid hydrogen/liquid oxygen engines (RL10A-3).

NASA awarded a study contract for the third stage (S-V) to Convair Astronautics on 21 October 1960. This stage (similar to Centaur) was designed for escape missions of the Saturn I vehicle. In March 1961 NASA reviewed the mission requirements and the development status of the entire Saturn program. Based on a decision that the primary mission of Saturn I would be to launch manned spacecraft into low earth orbits, development of the S-V stage was cancelled. However, Convair did supply dummy S-VD stages for the first four Saturn flight vehicles.

On 30 July 1962 Chrysler Corporation Space Division (CCSD) was awarded a contract for 21 booster stages, (S-I), which would be built at the Michoud Ordnance Plant near New Orleans, Louisiana. NASA subsequently restated the requirements for the number of boosters to be fabricated, checked out, tested and launched by CCSD.

On 20 August 1963 the contract was amended, reducing the number to 14. Two S-I stages were scheduled for the last two research and development vehicles; the remaining 12 were to be fabricated to the S-IB configuration.

Ten research and development vehicles constituted the Saturn I project. Three R&D and nine operational vehicles constituted the original Saturn IB program.

The letter "S" followed by the appropriate Roman numeral or letter(s), or both, identified the vehicle stages except for the payloads: S-I and S-IB (first stage), S-IV and S-IVB-200 (second stage), S-IVD (dummy version of the S-IV stage), S-IU (instrument unit), etc.

The Saturn I vehicles were of two basic designs: Block I (SA-1 through SA-4) consisted of live booster (S-I stage) and dummy upper stages (S-IVD, S-VD and payload); Block II (SA-5 through SA-10) consisted of an S-I stage, S-IV stage, instrument unit and payload. Each of the Saturn IB vehicles consisted of an S-IB stage, S-IVB-200 stage, instrument unit and payload.

The Saturn IB vehicle became the vehicle that launched the first manned Apollo mission (Apollo 7 in October 1968). After a period of storage three Saturn IB vehicles launched the three manned crews to the Skylab space station in 1973. The final Saturn IB launch took the first US crew to dock with a Russian spacecraft into orbit in 1975. In addition the final S-IVB stage to be produced was converted into the Orbital Workshop of the Skylab space station.

The Saturn I and Saturn IB vehicles opened the door for the successful flights of the larger Saturn V vehicle that took 12 men to the moon's surface and returned them safely to the earth between 1969 and 1972.

S-I-D (part of SA-D1 to SA-D4) Block I Dynamic Test Stage

Summary

Saturn I Dynamic Test Stage built and tested to simulate the dynamic environment of the first four Saturn I launch vehicles. Tested at MSFC in 1961 and 1962. Currently on display at MSFC.

Engines

The engine configuration currently installed in the stage (with locations not verified) is:

Position 101: H-1018
Position 102: H-1023
Position 103: H-1025
Position 104: H-1026
Position 105: H-1046
Position 106: H-1060
Position 107: H-5011
Position 108: H-5036

S-I-D stage being lowered into the dynamic stand at MSFC (9.6.1961)

Stage manufacturing

Construction of the Saturn I/IB Dynamic Test Tower at MSFC was started on 20 July 1960 and was completed on 17 April 1961 when the facility was handed over to the MSFC Test Division. In parallel the S-I-D booster was constructed at MSFC, Huntsville, starting on 7 March 1961. The booster was labeled, SA-D.

Engine gimbal tests on the S-I-D stage during April and May 1961 had indicated the advisability of increasing the stiffness of the engine control support structure in the booster. On 29 May 1961 MSFC performed a series of single-engine gimbal tests using the S-I-D stage to verify the effectiveness of the structural modifications to the engine support structure. As the test results were only marginally successful a new type of actuator servo valve was installed. Further tests were deemed satisfactory.

Stage testing

On completion of these tests the booster was moved to the newly-built Dynamic Test Stand in early June 1961. The S-I-D booster stage, configured to represent the S-I-1 to S-I-4 stages, was installed in the new tower for the first time on 9 June 1961. The dummy S-IV stage, manufactured by MSFC, dummy S-V stage, which had been delivered from Convair on 27 February 1961, and the payload body followed, forming the complete SA-D1 to SA-D4 vehicle. Within the tower the vehicle was suspended on huge springs.

Completion of the SA-D vehicle (6.1961) 6413237

S-I-D stage on display at MSFC (28.8.1962)

This was the first time this configuration had been assembled vertically. Although the test stand had been completed on 17 April 1961, in-house build up of the facility continued through August 1961.

The first dynamic tests were re-scheduled in order to add windscreens to the test tower and to reinforce the remaining three gimbaled engine positions. Dynamic testing began on 23 June 1961, under simulated lift-off conditions with de-ionized water loaded into the S-I-D, and dummy S-IV and S-V stages. The first tests were used to investigate the bending modes of the vehicle and also to continue studies into tank resonances.

Block I dynamic testing of the SA-D vehicle was concluded on 2 March 1962. However the SA-D vehicle remained in the test stand until May 1962 when it was removed. The dummy S-V stage was used later on a flight vehicle.

The block I Saturn booster stage became a horizontal exhibit at the MSFC Space Orientation Centre when it was dedicated on 29 August 1962. In December 1965 the stage was displayed vertically with dummy S-IV and S-V stages on top and the stage has remained in this position ever since. A survey by NASA in June 2004 confirmed the identification of this stage and recommended that the stage be categorized as a National Historic Monument.

Alan Lawrie with the SA-D vehicle at MSFC (2004)

S-I-T, S-I-T1 to S-I-T4.5 (SA-T, SA-T1 to SA-T4.5) Block I and Block II Static Test Stage

Summary

The first Saturn stage built (at ABMA) and the first Saturn stage test fired (also at ABMA). Used to simulate the Block I and Block II Saturn I booster stages and to verify the propulsion systems in static firings in those configurations. Currently on display at MSFC.

Assembly of the S-I-T booster in building 4705 at MSFC
(25.1.1960) 6000382

Engines

The engine configuration for static test SA-1 was possibly:

Position 101: No engine installed
Position 102: No engine installed
Position 103: No engine installed
Position 104: No engine installed
Position 105: No engine installed
Position 106: H-1007
Position 107: No engine installed
Position 108: H-1009

The engine configuration for static test SA-2 was possibly:

Position 101: No engine installed
Position 102: No engine installed
Position 103: No engine installed
Position 104: No engine installed
Position 105: H-1006
Position 106: H-1007
Position 107: H-1008
Position 108: H-1009

The engine configuration for static tests SA-3 to SA-28 (excluding SA-10) was possibly:

Position 101: H-1001
Position 102: H-1002
Position 103: H-1003
Position 104: H-1005
Position 105: H-1006
Position 106: H-1007
Position 107: H-1008
Position 108: H-1009

The engine configuration for static test SA-10 was possibly:

Position 101: No engine installed
Position 102: No engine installed
Position 103: No engine installed
Position 104: No engine installed
Position 105: No engine installed
Position 106: H-1007
Position 107: No engine installed
Position 108: H-1009

First Saturn S-I-T circumferential weld at MSFC building 4707 (15.5.1959) 5914021

Assembly of the S-I-T booster in building 4705 at MSFC
(1.2.1960) 6001841

S-I tail section mock up being installed in the MSFC stand
(4.1.1960)

The engine configuration for static tests SA-29 to SA-31 was possibly:

Position 101: H-5001
Position 102: H-5006
Position 103: H-5011
Position 104: H-5014
Position 105: H-2009
Position 106: H-2011
Position 107: H-2013
Position 108: H-2014

The engine configuration currently installed in the stage (with locations not verified) is:

S-I-T stage being transported to MSFC test stand (28.2.1960)
6002918

S-I-T stage being transported to MSFC test stand (28.2.1960)
6029131

Position 101: H-1001
Position 102: H-1002
Position 103: H-1003
Position 104: H-1005
Position 105: H-1006
Position 106: H-1007
Position 107: H-1008
Position 108: H-1009

Stage manufacturing

On 14 January 1959 ABMA began modifying the East side of the existing Static Test Tower in the East Test Area at MSFC for the static testing of Saturn I stages. The first prototype H-1 engine, H-1001, which would later be installed on the S-I-T stage, was delivered to ABMA in Huntsville on 28 April 1959. The first firing of this engine in the Power Plant Test Stand in

S-I-T stage being installed in the MSFC stand (28.2.1960)
6002782

S-I-T stage being installed in the MSFC stand (28.2.1960)
6002788

Huntsville occurred on 26 May 1959. The S-I-T stage became the first Saturn stage manufactured and assembled at the Army Ballistic Missile Agency in Huntsville. It was originally assembled without the engines installed, as these were to be inserted in the test stand.

Stage testing

On 4 January 1960 ABMA installed in the Static Test Tower East a mock-up of the tail section of the Saturn rocket for compatibility trials. On 1 February 1960 the

8th firing of the S-I-T booster (15.6.1960) 6009121

mock-up was removed, and some days afterwards, on 28 February, the S-I-T test stage, representing a test booster configuration, was lifted into the test stand. A 2.25 million-pound load test using eight hydraulic jacks was used to proof test the thrust structure. The proof load structure was removed and the suction ducts and engines for engine positions 106 and 108 were installed. Flak curtains were installed between the engines to prevent engine damage in the event of an engine explosion.

The first propellant load test was aborted when the LOX pump inlet reached a critical temperature. After de-tanking and adjustments the loading was performed successfully. After finalization of the electrical and instrumentation systems the technicians were ready to perform the first test firing.

The first firing of the test booster, S-I-T (SA-1), took place at 1058 CST on 28 March 1960. Only 2 engines were installed in this test that lasted 8 seconds. The test designation was SAT-01. For the second test two more engines were installed. This test, SAT-02, took place at 1115 CST on 6 April 1960 and was a four-engine test

One of the first 8 firings of the S-I-T booster (4-6.1960) 9808562

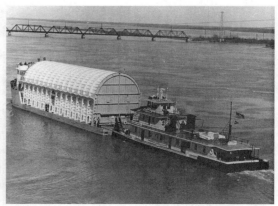

**Trial run of the Palaemon with the
S-I-T1 booster on board** (14.3.1961)

Alan Lawrie with the SA-T stage at MSFC (2004)

lasting 7 seconds. Prior to the next firing the remaining four engines were installed. The third firing, SAT-03, was also the first 8-engine test, and took place at 1728 CDT on 29 April 1960. The firing duration was 8 seconds. All further firings, except as noted, used all 8 H-1 engines.

Firing SAT-04 took place at 1732 CDT on 17 May 1960 and lasted 25 seconds. Firing SAT-05 took place at 1703 CDT on 26 May 1960 and lasted 35 seconds. Firing SAT-06 took place at 1656 CDT on 3 June 1960 and lasted 75 seconds. Firing SAT-07 took place at 1656 CDT on 8 June 1960 and lasted 110 seconds. Firing SAT-08 took place at 1709 CDT on 15 June 1960 and lasted 121 seconds. The stage was then removed from the tower and converted to simulate the SA-1 flight vehicle, including the installation of engine heat shields. On 1 July 1960 management of the Saturn project was transferred to MSFC.

The stage was returned to the tower in the SA-T1 configuration. Firing SAT-09, with 8 engines, took place at 1801 CST on 2 December 1960 and was aborted after only 1.7 seconds. The reason for the abort was that the low ambient temperatures had caused the gas generator fuel check valves to be sluggish in opening. This in turn had resulted in a LOX-rich mixture that had raised the temperature in the turbines and had caused all eight

sets of turbine blades to become eroded. After this problem the next firing, SAT-10, was performed with only 2 engines, and took place at 1449 CST on 10 December 1960 with a duration of 6 seconds. Firing SAT-11, with 8 engines, took place at 1639 CST on 20 December 1960 with a duration of 61 seconds.

Firing SAT-12, also in the SA-T1 configuration, took place at 1648 CST on 31 January 1961, with a duration of 113 seconds. LOX gauge measurements showed zero level at cut-off and the hydraulic system appeared to function properly. The next firing, SAT-13 was conducted at 1648 CST on 14 February 1961, with a duration of 109 seconds. The LOX and fuel pressurizing systems were operated for the first time according to the sequence and appeared to function correctly. This was the last of the second series of static firings.

The SA-T1 booster was removed from the Static Test Tower East at MSFC on 2 March 1961 in preparation for acceptance testing of the S-I-1 stage. The SA-T1 stage was transported to the barge Palaemon for a "shakedown cruise" practice trip with the new barge. The practice trip was conducted to acquaint the crew with the handling of the barge and to permit studies of the stresses to which the rocket would be subjected to during its journey to Cape Canaveral. The Palaemon left the MSFC dock on 14 March on its first training trip down the Tennessee River, carrying the SA-T1 stage (S-I-T1). After a three-day, 225 mile trip along the Tennessee River, the barge Palaemon returned to the Redstone Arsenal Dock (MSFC) on 17 March.

Alan Lawrie with the SA-T stage at MSFC (2004)

On 20 March the SA-T1 booster was moved from the barge to the Quality Division, building 4708, at MSFC for inspection and modification to the SA-T2 configuration, representing the SA-2 launcher.

All engines were removed and the LOX domes were replaced with reheat-treated domes. After being modified to the SA-T2 designation the test booster was installed in the Static Test Tower East on 27 May 1961.

The first firing in the SA-T2 configuration (SAT-14) took place at 1551 CDT on 27 June 1961, with a duration of 29.9 seconds. Cut-off was by manual command, predicated upon the completion of the scheduled special engine gimbal program. Major test objectives were to evaluate the effectiveness of the new actuator servo valve and the stiffening of the control engine thrust structure, and to check vehicle performance after build-up to the SA-2 configuration.

The second firing in the SA-T2 configuration (SAT-15) took place at 1620 CDT on 7 July 1961, with a duration of 118 seconds. The third firing in the SA-T2 configuration (SAT-16) took place at 1441 CDT on 18 July 1961, with a duration of 116 seconds. These tests evaluated modifications to reduce engine structure vibration, evaluated flame curtain materials, and checked out a LOX depletion system similar to that used on SA-1. During the third firing MSFC simulated for the first time the in-flight engine cut-off sequence, which was, shut down of the in-board engines 6 seconds before shut down of the out-board engines.

Three further firings were performed during August in the SA-T2 configuration. Firing SAT-17 took place at 1709 CDT on 3 August 1961 but was aborted after only 1.2 seconds of a planned 114 second firing when instrumentation indicated an unacceptably high temperature of the LOX pump inlet on engine # 101. Firing SAT-18 took place at 1543 CDT on 7 August 1961 with a duration of 123 seconds. Finally, firing SAT-19 took place at 1542 CDT on 25 August 1961 with a duration of 114 seconds. The S-I-T2 stage was removed from the Static Test Tower East on 6 September 1961 and converted to the SA-T3 configuration, representing the SA-3 vehicle.

After re-installation in the test stand the first firing in the SA-T3 configuration, SAT-20, took place at 1701 CST on 30 November 1961. This test was conducted to investigate flight cut-off sequencing, perform an "engine-out" test, and study fuel and LOX tank levels. The test was prematurely terminated after 95 seconds by the automatic fire detection system, although no damage occurred. The second firing in the SA-T3 configuration, SAT-21, occurred at 1542 CST on 19 December 1961 and lasted for 68 seconds. The third firing, SAT-22, took place at 1639 CST on 18 January

1962 and lasted for 122 seconds. The fourth firing, SAT-23, took place at 1641 CST on 6 February 1962 and lasted 46 seconds. The fifth, and final, firing in the SA-T3 configuration, SAT-24, took place at 1640 CST on 20 February 1962. This test was scheduled to last until LOX depletion cut-off, but was terminated after 55 seconds due to a fire indication at engine number 106. No damage was found during the post-test examination. The stage was then converted to S-I-T4, representing the SA-4 vehicle.

The first of four static firings in the SA-T4 configuration, SAT-25, took place at 1646 CDT on 19 June 1962 for a duration of 32 seconds. The second firing took place at MSFC at 1640 CDT on 12 July 1962. That firing (SAT-26) was terminated after 12.04 seconds because lube oil pressure in the turbo-pump on engine # 103 went below the redline value. The pressure reading was later found to be erroneous because of a broken lead wire. At 1656 CDT on 13 July 1962 the third static firing of the SA-T4 (SAT-27) took place. The firing was aborted after 20.43 seconds when a loose connection in the instrumentation lead wires caused the fuel inlet pressure to engine # 108 to read "zero". The fourth test of the SA-T4 configuration of the vehicle (SA-28) took place at 1640 CDT on 17 July 1962. This 120.00 seconds full duration test was fully successful. The stage was removed from the Static Test Tower on 20 July 1962.

The vehicle was then upgraded to the SA-T4.5 configuration, which included replacing the Block I H-1 engines (165,000 pounds thrust) with Block II engines (188,000 pounds thrust) and changing the structure slightly to the Block II SA-5 configuration. The SA-T4.5 was re-installed in the Static Test Tower in mid October.

The firings in the SA-T4.5 configuration were, SAT-29, conducted at 1651 CST on 26 October 1962 with a duration of 31.00 seconds; SAT-30, conducted at 1644 CST on 2 November 1962 with a duration of 65.00 seconds; SAT-31, conducted at 1642 CST on 9 November 1962 with a duration of 115.00 seconds. During the firings technicians checked the integrity of the propulsion system and the effect of the up-rated engines on the flame deflector.

After the tests were successfully concluded the stage was removed from the test stand on 15 November 1962 and shipped to Michoud for use in checking out facilities. Later, possibly in July 1964, the stage was returned to MSFC and eventually placed horizontally next to the Static Test Tower in the East Test Area. The stage has remained there ever since, exposed to the elements. A survey by NASA in June 2004 confirmed the likely identification of this stage and recommended that it be submitted as a National Historic Monument.

S-I-1

Summary

First Saturn flight booster stage. Assembled with 165,000 pounds thrust H-1 engines. Had to be transported around Wheeler Dam after its collapse. Built and tested at MSFC and launched from Cape Canaveral with dummy upper stages. Sub-orbital test flight.

Engines

The engine configuration in the stage for static testing and for launch was:

Position 101: H-1016
Position 102: H-1017
Position 103: H-1019
Position 104: H-1021
Position 105: H-1011
Position 106: H-1012
Position 107: H-1013
Position 108: H-1015

Stage manufacturing

Assembly of the booster stage for the first Saturn flight vehicle, S-I-1, began in Huntsville on 26 May 1960. When assembly of the S-I-1 booster stage was completed it was delivered to functional checkout at MSFC on 16 January 1961. The stage would ultimately fly with a dummy S-IV stage, manufacture of which was completed by MSFC in January 1961, and a dummy S-V stage, which was delivered from Convair Astronautics to MSFC on 8 February 1961. During February 1961 the S-I-1 stage was temporarily assembled together with the dummy S-IV stage, dummy S-V stage and payload cover in a horizontal orientation to form the first assembled Saturn rocket.

S-I-1 stage in MSFC building 4705 (1.1961) 6100433

Stage testing

The stage was installed in the Static Test Tower East at MSFC on 6 March 1961. The first firing of this stage was delayed two weeks, during April, in order to permit decontamination of the LOX system. After this was achieved the first firing (SA-01) was performed at 1638 CDT on 29 April 1961. The test duration was 30 seconds. Cut-off was given by a timer at a pre-determined setting.

The second firing (SA-02) was conducted at 1612 CDT on 5 May 1961, the same day as Alan Shepard's first Mercury flight. The firing duration was 44.17 seconds. Although scheduled as a long-duration firing, the test was prematurely terminated by the vehicle fire detection system because of a leaking pressure pickup in a gas generator. The pickup was omitted from all engines for subsequent firings.

The third and final static firing of the stage (SA-03)

S-I-1 stage in MSFC building 4705 with S-I-2 stage at rear
(1.1961) 6413240

S-I-1 stage in MSFC building 4705 (18.1.1961) 6412470

Trial assembly of the SA-1 stages at MSFC (15.2.1961) 6521237

took place at 1548 CDT on 11 May 1961. The test was successful, with a duration of 111 seconds. At the completion of this test final repair and assembly work began on the stage.

On 25 May 1961 the stage was removed from the test stand, after being there for 12 weeks. Following modification and repair the booster was transferred to final checkout on 12 June 1961, in preparation for shipment to Cape Canaveral. Modification and repair continued during this checkout. At the same time MSFC tested the

Trial assembly of the SA-1 stages at MSFC (15.2.1961) 6521238

Installing the S-I-1 stage in the MSFC stand (6.3.1961) 6518198

S-I-1 stage in MSFC stand (3.1961) 6110711

SA-1 dummy payload being moved around Wheeler Dam
(5.8.1961)

dummy S-IV stage for the SA-1 flight (20 to 25 May) and continued preparations for shipping the stages to Cape Canaveral.

The dummy S-V stage was the first to be transported to Cape Canaveral. The barge Palaemon embarked on its first trial run to the Cape, leaving Huntsville on 17 April 1961. It carried as cargo the S-V dummy stage for the SA-1 flight and ballast simulating the S-I-1 stage. However, it was hit on the port side by a Norwegian tanker on 21 April and the damage necessitated that the Palaemon had to enter a New Orleans shipyard for repairs. These were completed on 25 April and the barge continued on to the Cape. After arriving successfully on 1 May and off-loading the S-V dummy, the Palaemon departed two days later for the return journey to Huntsville, where it arrived at 0600 CDT on 15 May. The S-V dummy stage was transferred to Hangar D at Cape Canaveral.

However, on 2 June 1961 a lock at the Wheeler Dam, just downstream of Huntsville, collapsed halting all movement on the Tennessee River and trapping the barge Palaemon upstream of the blockage. MSFC decided to transport the SA-1 stages from Huntsville to the Wheeler Dam using the Palaemon. At the Wheeler Dam the stages were off-loaded and transported by land to a point downstream of the blockage where they could be loaded onto another barge. In support of this plan MSFC obtained a Navy barge which had been moth-balled at Pensacola, Florida. Emergency modifications were made to the barge so that it could carry the SA-1 stages. The barge was renamed Compromise.

In Huntsville final acceptance testing of the S-I-1 stage began on 12 June 1961. The first operation accomplished was the mechanical mating of the S-IV dummy stage. Checkout of the S-I-1 stage continued until it was successfully completed in early August.

On 5 August 1961 the S-I-1 stage, dummy S-IV stage

and dummy payload were loaded on the Palaemon and transported from Huntsville to the Wheeler Dam. After the transfer around the Dam to the Compromise, the rocket stages arrived at Cape Canaveral on 15 August 1961. The booster stage was transferred to Complex 34 and the other stages were moved to Hangar D.

The S-I-1 booster was erected on the launch pedestal on 20 August 1961. On 23 August the dummy S-IV, dummy S-V and dummy payload were mounted on top of the booster. The service structure was removed on 15 September. On 27 September the fuel test was completed and on 4 October the LOX loading test was performed. The engine swivel checks were performed on 13 October. Simulated flight tests were performed on 16 and 23 October. The fuel was loaded on 26 October and the LOX loaded on the flight day.

The S-I-1 stage formed the first stage of the SA-1 vehicle that was launched from Launch Complex 34 at 1006:03.89 EST on 27 October 1961.

S-I-1 stage being offloaded from the Palaemon at Wheeler Dam (5.8.1961)

S-I-2

Summary

Second Block I Saturn booster. Manufactured and tested at MSFC. Transported around Wheeler Dam because of the collapse of the lock. Launched from Cape Canaveral on sub-orbital test flight. Included Project Highwater, high altitude release of water.

Engines

The engine configuration in the stage for static testing and for launch was:

Position 101: H-1032
Position 102: H-1033
Position 103: H-1034
Position 104: H-1038
Position 105: H-1028
Position 106: H-1029
Position 107: H-1035
Position 108: H-1036

Stage manufacturing

Fabrication of the tanks for the second Saturn flight booster was completed during December 1960 and assembly of the booster began on 27 December. However, in early 1961 assembly of the S-I-2 booster was approximately 8 weeks behind schedule, due to late delivery of components, modification of engines, and design changes necessitated by static firing results. The dummy S-IV stage for the SA-2 vehicle, which was manufactured by MSFC, was structurally complete by the middle of 1961. The dummy S-V upper stage was received from Convair in June 1961. The S-I-2 booster was transferred from the assembly area to the checkout at MSFC on 1 August 1961.

Stage testing

The S-I-2 stage was placed in the Static Test Tower East in early October and static fired at 1659 CDT on 10 October 1961 for a duration of 32 seconds. The firing was designated SA-04. The purpose of the test was to check the reliability and performance of the booster and gimbal systems. A second, and final, firing of this stage (SA-05) took place at 1641 CST on 24 October 1961 for a duration of 119 seconds. Test objectives included evaluation of the flight cut-off sequence.

The S-I-2 booster stage was removed from the test stand on 2 November 1961, and returned to the Manufacturing Engineering Division for modifications and repair.

On 16 February 1962 the dummy S-IV stage, the dummy S-V stage and the payload body were loaded aboard the barge Palaemon at MSFC dock and transported to Wheeler Dam the next day. The Palaemon returned to pick up the S-I-2 stage and brought it to Wheeler Dam on 17 February. As with the SA-1 vehicle, the cargo had to be transferred by land around the Dam on a specially manufactured 1 mile roadway, following the earlier collapse of the Dam lock. The transfer took about 1 hour. The stages were loaded aboard the next barge, Promise (formerly named Compromise), which was used to transport the stages to Cape Canaveral, where they arrived on 27 February 1962. Modifications to the barge Compromise had been completed on 14 December 1961 and the barge had been re-named Promise to support the SA-2 and future operations. Following the delivery, and repair to the lock, Promise arrived at the MSFC dock for the first time on 18 May 1962 in readiness for transporting the SA-3 vehicle all the way to Cape Canaveral.

The S-I-2 stage was transferred to launch complex 34 on the day of arrival, 27 February. The SA-2 booster was erected on Launch Pad 34 on 1 March 1962. The same day the upper stages and dummy payload were transferred to pad 34 and erected on top of the booster the following day. The service structure was removed on 27 March. On 9 April the fuel test was performed and on 12 April the LOX loading test was performed. The simulated flight test was performed on 19 April. RP-1 fuel was loaded into the vehicle on 23 April, and LOX was loaded on the launch day. The S-I-2 booster stage formed the first stage of the SA-2 vehicle that was launched at 0900:34.41 EST on 25 April 1962.

S-I-3

Summary

Third Block I Saturn booster stage. First Saturn booster to be transported all the way to Cape Canaveral on one barge. Sub-orbital test flight and second Project Highwater mission.

Engines

The engine configuration in the stage for static testing and for launch was:

Position 101: H-1045
Position 102: H-1047
Position 103: H-1048
Position 104: H-1049
Position 105: H-1040
Position 106: H-1041
Position 107: H-1042
Position 108: H-1043

Stage manufacturing

Fabrication of the LOX and fuel tanks for the S-I-3 stage started in April 1961. Assembly of the stage began on 31 July 1961 and was completed on 26 December 1961. On 8 January 1962 the S-I-3 stage successfully completed functional and engine pressure tests and entered the pre-static checkout phase. On 19 March 1962 the S-I-3 stage completed pre-static pressure tests and instrumentation and component functional tests. The stage was installed in the Static Test Tower East at MSFC later the same day.

Stage testing

The first static firing of the S-I-3 stage, SA-06, took place at 1713 CDT on 10 April 1962 for a duration of 31 seconds. During the firing a defective bearing in one of the turbo-pumps was discovered. This manifested itself as excessive turbo-pump vibration and meant that the engine had to be removed and defective parts replaced before the test could be repeated. The engine was recertified and returned to the stage and the second repeat short firing, SA-07, took place at 1642 CDT on 17 May 1962 for a duration of 31.02 seconds. Overall performance was excellent with the exception of a hang-fire ignition of the solid propellant gas generator at position # 106. The third, and final, firing of the S-I-3 booster, SA-08, took place at 1642 CDT on 24 May 1962 for a duration of 119.43 seconds.

The S-I-3 stage was returned to the Quality Assurance Division in late May 1962 to clear the Static Test Tower East for the SA-T4 stage.

The S-I-3 stage, together with dummy S-IV and S-V stages, were loaded aboard the barge Promise, which departed MSFC on 9 September 1962. As the lock at the Wheeler Dam had been repaired on 18 April 1962 the barge was able to make the complete journey without need for intermediate road transportation, as had happened with the first two flight deliveries to Cape Canaveral. However, the journey was about two weeks later than planned. It followed the Tennessee River to the Ohio at Paducah, Kentucky; the Ohio to the Mississippi at Cairo, Illinois; proceeded down the Mississippi to the Gulf of Mexico; and then went around the tip of Florida, arriving at Cape Canaveral on 19 September 1962.

The S-I-3 stage was erected on Launch Complex 34 on 21 September 1962, followed by the dummy upper stages and the payload body on 24 September. The service structure was removed on 19 October, and on 31 October the fuel test was completed and the dummy upper stages were loaded with water. On 2 November the LOX loading test was completed and on 9 November the retro rockets were installed. The simulated flight test was performed on 13 November and the following day the fuel was loaded.

The S-I-3 stage formed the first stage of the SA-3 vehicle that was launched from Launch Complex 34 at 1245:02 EST on 16 November 1962.

The barge Promise, with SA-3 on board, passes through a Tennessee River lock (9.9.1962)

S-I-4

Summary

Fourth and final Block I Saturn booster stage. The short stage firing at MSFC was witnessed by the President and Vice President. Launched from Cape Canaveral. Included an engine-out test in flight.

Engines

The engine configuration in the stage for static testing and for launch was:

Position 101: H-1054
Position 102: H-1055
Position 103: H-1056
Position 104: H-1057
Position 105: H-1051
Position 106: H-1052
Position 107: H-1053
Position 108: H-1058

Stage manufacturing

Manufacturing activities on the S-I-4 vehicle started on 31 July 1961, assembly began on 2 January 1962 and was completed on 26 May 1962. Pre-static checkout of the S-I-4 booster was completed in July 1962.

Stage testing

The stage was erected in the Static Test Tower East at MSFC on 1 August in preparation for the static firings. After the propellant loading test, conducted on 16 August, technicians found foreign material in the LOX pump inlet screens and decided to flush the LOX system prior to the first firing. On 18 August it was discovered that several areas of the stage were contaminated by oil. This required nearly 3 weeks of cleaning to resolve the issue. The LOX system was finally purged on 5 and 6 September in preparation for the static firing.

The first firing, SA-09, was conducted at 1117 CDT on 11 September 1962 with a duration of 31.53 seconds. Afterwards technicians again discovered contamination in the facility and stage GN2 systems and had to clean these systems a second time. This test firing was somewhat unique as it featured the only Saturn I/IB flight booster stage to be test fired in the morning. The reason for the early time of the firing was that the firing was witnessed by President John F Kennedy and Vice President Lyndon B Johnson, who were visiting MSFC that day and the firing time had to be adjusted to their schedule.

The second firing of the S-I-4 stage, SA-10, took place at 1642 CDT on 26 September 1962 with a duration of 121.5 seconds (inboard engines) and 127.43 seconds (outboard engines).

The stage was removed from the Static Test Tower on 1 October and delivered to the Manufacturing Engineering (ME) Division of MSFC. Following post static modifications the stage was transferred to the Quality Division on 22 October 1962, for final checks.

During the first two weeks of January 1963 QA Division at MSFC completed the post static checkout of the stage and ME Division completed structural modifications to the dummy S-IV stage.

On 20 January 1963 the S-I-4 booster stage, the dummy S-IV stage, the dummy S-V stage, the payload and the water ballast tank were shipped to Cape Canaveral on board the barge Promise. The barge arrived at Cape Canaveral on 2 February. The S-I-4 stage was erected on pad 34 on 4 February, followed by the dummy upper stages on the following day.

The S-I-4 stage formed the booster stage for the SA-4 vehicle that was launched at 1511:55 EST on 28 March 1963.

President Kennedy is met at MSFC by Dr von Braun prior to the S-I-4 firing (11.9.1962) 9806978

S-I-4, S-I-6 and S-I-7 in assembly in building 4705 at MSFC (13.1.1963) 6413388

S-I-5

Summary

First Block II S-I stage. Incorporated up-rated 188,000 pounds thrust H-1 engines, elongated propellant tanks and the addition of 8 fixed fins. Launched from Cape Kennedy with a live S-IV upper stage. Orbital test flight.

Engines

The initial engine configuration in the stage was:

Position 101: H-5002
Position 102: H-5003
Position 103: H-5004
Position 104: H-5006 – removed from the stage and replaced by H-5005
Position 105: H-2001
Position 106: H-2002
Position 107: H-2003
Position 108: H-2004

Spare engine: H-5005 – installed in place of H-5006

The engine configuration in the stage for static testing and for launch was:

Position 101: H-5002
Position 102: H-5003
Position 103: H-5004
Position 104: H-5005
Position 105: H-2001
Position 106: H-2002
Position 107: H-2003
Position 108: H-2004

Stage manufacturing

All H-1 engines were received at MSFC from Rocketdyne in April 1962. On 7 July 1962 MSFC began the assembly of the S-I-5 booster stage in Huntsville. MSFC returned the first shipment of S-I-5 booster propellant tanks to Chance Vought because of deficiencies. The Center received and accepted the second set of tanks in July 1962. The ME division completed the connection of the tanks to the tail assembly on 31 July 1962. In August work to up-rate the engines to 188,000 pounds thrust was completed. The four H-1 inboard engines were installed in the stage in August, but it was not until October before the outboard engines could be installed, the final one being in place on 11 October. The delay was due to the late arrival of the installation equipment. On 6 November 1962 the stage was released to the Test Division at MSFC for pre-static checkout. However, the vehicle was only partially assembled at this stage and the completion of the work of installation of components was completed by the Manufacturing Engineering Division after the transfer.

Stage testing

The MSFC Test Division employed a double shift in its preparations for the static firing of the S-I-5 stage. Early in January 1963 the ME Division completed the installation of all components in the S-I-5 booster while the QA Division completed the pre-static checkout of the stage. The stage was installed in the Static Test Tower East on 28 January 1963.

On 13 February during a propellant loading test technicians noted excessive leakage on the bottom manhole cover of the center LOX tank. The propellant loading test was stopped, the tanks emptied of propellant, the gasket to the center tank replaced, and a higher torque applied to the bolts. The propellant loading test was repeated successfully the following day, although some leaks were noted. As a result the center manhole cover was removed and returned to the ME Division for modification.

On 15 February MSFC replaced engine # 104 (H-5006)

which was thought to be faulty. The engine was returned to Rocketdyne because of suspected damage to the turbo-pump. However, subsequent inspection of all turbo-pump components revealed that the pump was in satisfactory condition. The engine was re-built and became a spare for the S-I-5 and S-I-6 stages. The engine is currently on permanent loan to the University of Utah. It was replaced with engine H-5005.

On 25 February the centre manhole cover was re-installed and the vehicle was successfully tanked with propellant on 27 February.

First firing of the stage (SA-11) was conducted in the MSFC Static Test Tower East at 1647 CST on 27 February 1963, with a duration of 31.96 seconds.

The full duration firing, SA-12, took place at 1617 CST on 13 March 1963 with a duration of 144.44 seconds. Due to anomalies in the propulsion system during this test it was decided to repeat the firing. The repeat firing (SA-13) was started at 1640 CST on 27 March 1963 and had a duration of 143.47 seconds. The stage was removed from the test stand on 2 April 1963. During rework and post static checkout of the S-I-5 stage other discrepancies in the propulsion system were discovered

Erection of the S-I-5 booster at KSC (23.8.1963)

and the necessary corrections were made.

In mid-June post static checkout of the stage was completed. After completion of the S-IU-5 acceptance tests, both the S-I-5 and the S-IU-5 were electrically mated and programmed through an automatic launch sequence. Test results verified acceptability of the S-I-5 and S-IU-5 for launch preparations. In mid July the S-I-5 and S-IU-5 were prepared for shipment. They were briefly placed in storage, awaiting the availability of the S-IV-5 second stage.

The S-I-5 stage, together with the S-IU-5 and SA-5 payload, were transported from MSFC to Cape Canaveral on board the barge Promise, leaving Huntsville on 11 August 1963 and arriving at Cape Canaveral on 21 August. During the period 23 to 26 August the S-I-5 stage was erected on Launch Complex 37B. The booster fins were attached, followed by an S-IV spacer atop the booster, in lieu of the actual S-IV-5 second stage which was still in Sacramento. On 5 September the S-IU-5 was installed on top of the dummy second stage.

With the arrival of the real S-IV-5 second stage on 21 September, the dummy S-IV stage was removed from the stack and the real vehicle stacked on 11 October 1963.

During November 1963 the launch of SA-5 was postponed due to the discovery of cracks in fuel line fittings in the S-I-5 stage during pressure checks conducted as a part of the pre-launch checkout. The fittings would be replaced on this and all subsequent stages.

The first launch attempt on 27 January 1964 was aborted because a test flange was inadvertently left in the S-I-5 stage LOX replenish line, preventing the flow of LOX to replenish the stage. The S-I-5 stage formed the booster stage of the SA-5 vehicle which was launched at 1125:01 EST on 29 January 1964 from LC-37B.

S-I-6

Summary

Second Block II Saturn I booster stage. Launched from Cape Kennedy with a live S-IV stage and Apollo Boilerplate capsule on top. One of the first stage H-1 engines failed in flight.

Engines

The engine configuration in the stage for the initial build and for the first static testing was:

Position 101: H-5007

Position 102: H-5008
Position 103: H-5009
Position 104: H-5010
Position 105: H-2005
Position 106: H-2006
Position 107: H-2008
Position 108: H-2009 - subsequently removed from stage and replaced by H-2007

Spare engine: H-2007 – installed in place of H-2009

The engine configuration in the stage for the second static firing and for launch was:

Position 101: H-5007
Position 102: H-5008
Position 103: H-5009
Position 104: H-5010
Position 105: H-2005
Position 106: H-2006
Position 107: H-2008
Position 108: H-2007

Stage manufacturing

MSFC began assembling the S-I-6 booster stage on 25 September 1962. Progress was delayed by a lack of vendor supplied items. On 21 January 1963 the QA Division at MSFC began pre-static checkout of the stage. During this period a number of parts were installed that had been unavailable during assembly. Pre-static checkout was completed on 3 April.

Stage testing

The S-I-6 stage was installed in the Static Test Tower East on 22 April 1963. The propellant loading and subsystem test was performed on 7 May. The first firing, SA-14, was performed at 1640 CDT on 15 May 1963. At the conclusion of the 33.75 second firing the main LOX valve of engine # 108 failed to close at cut-off. The resulting fire severely damaged the engine. Technicians removed the propellants from the stage by using the fuel and LOX emergency dump system. Inspection revealed that the chamber tubes and injector baffles were partially melted. On 20 May engine # 108 (H-2009) was replaced with a spare (H-2007) and on 28 May all the main LOX valves were removed, modified and replaced on the stage on 31 May. The removed engine (H-2009) was eventually transferred to the Smithsonian.

The second and final firing, SA-15, took place at 1642 CDT on 6 June 1963. The duration of this successful test was 142.37 seconds. The inboard engines were cut off by LOX low level sensors after 136 seconds and the outboard engines continued to fire to the conclusion of the test. The test was successful except that the thrust level of engine # 105 (H-2005) was above specification limits, despite the fact that the engine had been re-orificed after the first test firing during which it also exhibited a similar high thrust rating. Engine # 108 (H-2007) also exhibited a higher than specified thrust of 194,841 pounds. The stage was removed from the test stand on 17 June and it was returned to the ME Division for post-static modification.

On 18 June MSFC began post-static modification and rework of the stage. Some bent and broken anti-slosh baffles in the 70-inch LOX tanks were discovered and repaired. To investigate the cause of this problem MSFC decided to install strain gauges in the S-I-7 stage during its static tests.

Final checkout of the S-I-6 booster stage at MSFC was initiated in July 1963. Pressure and functional checks were completed by 26 July. In August the stage underwent weight and alignment tests, and in early October performance testing of the stage was initiated. The Laboratory performed approximately 55% of the tests with automated checkout equipment. The test program was successfully completed on 25 October and the stage was accepted for shipment to the launch site; however due to the excessive number of cracked sleeves found on flared tubing fittings of S-I-5 at the launch site it was decided to replace all critical flared tubing on S-I-6.

In January 1964 MSFC completed replacement of the critical tubing assemblies, completed mechanical and electrical checkout of the stage and prepared the stage for shipment to KSC.

The S-I-6 stage, together with the S-IU-6, was shipped on board the barge Promise from Huntsville on 7 February 1964, arriving at Cape Kennedy on 18 February. They were unloaded the following day. The S-I-6 stage was erected on Launch Pad 37B on 20 February and seven days later the large fins were attached.

The S-IV-6 stage, which arrived on 22 February, was mated with the S-I-6 stage on 19 March. The IU was stacked atop the S-IV stage on 23 March. The Apollo spacecraft, BP-13, comprising a CSM, adapter and LES, was placed on top the stack on 2 April 1964. The RP-1 fuel loading test took place on 6 April.

During propellant loading tests on 11 May the propellant utilization system malfunctioned due to failure of the LOX sump screen mesh in the S-IV-6 LOX tank. A reinforced screen was added as a repair. The simulated flight test took place on 20 May and the flight RP-1 fuel was loaded on 23 May. The planned launch on 26 May was scrubbed due to an environmental control system compressor malfunction in the ground facilities equipment. Launch of the SA-6 vehicle, with the S-I-6 boost-

er stage, occurred at 1207:00 EDT on 28 May 1964. During the flight engine # 8 (H-2007) cut off after 117.28 seconds, some 23 seconds earlier than planned. The cause was determined to be a malfunction of the turbo-pump. It was believed that the most probable cause of the malfunction was failure of the A-B gear mesh due to root fatigue of the B gear.

S-I-7

Summary

Penultimate Block II Saturn I stage to be manufactured at MSFC. Launched from Cape Kennedy.

Engines

The engine configuration in the stage for the first static firing was:

Position 101: H-5013
Position 102: H-5014 – subsequently removed from the stage and replaced by H-5027
Position 103: H-5015
Position 104: H-5016
Position 105: H-2010
Position 106: H-2011
Position 107: H-2012
Position 108: H-2015

Spare engine: H-5027 – installed in place of H-5014
Spare engine: H-5022 – (not used, transferred to S-I-9 spare, transferred to Atlas program)

The engine configuration in the stage for the second static firing was:

Position 101: H-5013
Position 102: H-5027
Position 103: H-5015
Position 104: H-5016
Position 105: H-2010
Position 106: H-2011 – subsequently removed from the stage and replaced by H-2019
Position 107: H-2012
Position 108: H-2015 – subsequently removed from the stage and replaced by H-2021

Spare engine: H-2019 – installed in place of H-2011
Spare engine: H-2021 – installed in place of H-2015

The engine configuration in the stage for launch was:

Position 101: H-5013
Position 102: H-5027
Position 103: H-5015
Position 104: H-5016

Position 105: H-2010
Position 106: H-2019
Position 107: H-2012
Position 108: H-2021

Stage manufacturing

The H-1 engines were received from Rocketdyne in August 1962. On 7 January 1963 MSFC began clustering the tanks for the S-I-7 stage in Huntsville. The stage assembly was completed on schedule and it was released for pre-static checkout on 25 May. In June MSFC completed the status and continuity checks and instrumented the fuel and LOX tank baffles to determine the cause of the bending and fracture seen during the S-I-6 static firings. Performance testing of the stage was started on 13 June 1963.

In July 1963 MSFC's Manufacturing Engineering (ME) Laboratory modified all 8 H-1 engines for the S-I-7 stage. The gas generator LOX orifice was also increased to compensate for an expected loss in engine performance. The Quality Laboratory then resumed the pre-static checkout of the stage. The pre-static test program was completed satisfactorily on 22 August 1963 and the stage was released to the MSFC Test Laboratory for static firing.

Stage testing

Booster stage S-I-7 was erected in the Static Test Tower East at MSFC on 6 September 1963. In preparation for static firing a propellant loading test was performed on 24 September 1963 followed by a fuel tank pressure test on 30 September.

Static firing SA-16 was initiated at 1638 CDT on 2 October 1963 for the planned duration of 33.78 seconds. Following this firing engine # 102 (H-5014) was removed because of excessive thrust chamber tube leakage and a spare H-1 engine (H-5027) was installed in its place. The remaining 7 engines were also re-orificed because they exhibited higher than specified thrust. Meanwhile engine H-5014 became a spare, was turned over to R&D and was eventually scrapped.

Test SA-17 was performed at 1639 CST on 22 October 1963 with a duration of 138.93 seconds (inboard engines), 145 seconds (outboard engines). After this firing external fuel tube leakage was observed at engine # 108 (H-2015) and it was decided to replace the engine with a spare (H-2021) prior to launch. Meanwhile engine H-2015 became a Test Division spare and was eventually transferred to the Smithsonian.

After the firings the stage was removed from the test stand on 4 November 1963 and returned to the ME Lab-

oratory for repair and post static checkout On 26 November 1963 the stage was moved to the Quality Laboratory for final checkout. Alignment checks were completed and pressure testing had begun when, in late December, the final checkout was discontinued until tubing could be replaced, following the problem on SA-5.

At MSFC, replacement of critical tubing assemblies was completed on 10 February 1964, and the vehicle was returned to checkout on 14 February. Post static checkout was completed on 12 May.

On 28 May the S-I-7 and S-IU-7 left Huntsville on board the barge Promise, which arrived at Cape Kennedy on 7 June 1964. The SM and adapter arrived via aircraft the same day. The S-I-7 was erected on Launch Pad 37B on 9 June 1964. The S-IV-7 stage arrived via the Pregnant Guppy on 12 June 1964. The Command Module, BP-15, arrived via aircraft on 15 June. The S-IV-7 stage was erected atop the booster stage on 19 June. The IU was temporarily erected for drill marking on 19 June and finally installed on 22 June. Power was applied to the S-IV stage on 23 June and the spacecraft was erected three days later.

Also on 26 June a 3-inch crack in the LOX dome on one of the S-I-7 H-1 engines, position # 106 (H-2011), was

Crack in LOX dome on S-I-7 engine H-2011 (6.1964)

detected. The discovery led to the removal of the engine and shipment to MSFC for inspection. This engine became a Test Division spare and was eventually transferred to the Smithsonian. The engine was replaced in the S-I-7 stage with engine H-2019. It was determined that stress corrosion had caused the cracks. It was decided to replace the LOX domes on all the S-I-7 stage H-1 engines and in the period 4 July to 20 July all the engines were returned to Rocketdyne's Neosho, Missouri plant to be retrofitted with new domes. The investigation into the stress corrosion cracking resulted in a change of material in the manufacture of LOX domes from 7079-T6 aluminum forging to 7075-T73 aluminum alloy. The part was made by the Alcoa Vernon Plant, Vernon, California.

Checkout of the refurbished engines was achieved by 31 July. On 4 August the S-I-7 and S-IV-7 stages underwent full pressure checks. The spacecraft LES was erected on 17 August and the all systems overall vehicle systems test performed two days later. On 27 August Hurricane Cleo passed the area and the launch complex was secured. On 3 September a simulated flight test revealed a LES bolt failure due to stress corrosion. The tower was removed to a remote area. The following day all LES bolts were replaced and the LES re-installed.

On 9 September Hurricane Dora passed the area and once again the launch complex was secured. Fuel loading occurred on 12 September and the Countdown Demonstration Test took place on 14 and 15 September. Following the CDDT, a crack in the weld that held two LOX tank vent lines to the skin of the S-I-7 stage was detected. As a repair doublers were riveted to the attachment to provide additional support and prevent complete rupture of the weld during the launch. Booster stage LOX loading and S-IV fuel and oxidizer loading took place on the day of launch.

The S-I-7 stage formed the first stage of the SA-7 vehicle that was launched from Launch Complex 37B, Cape Kennedy, at 1122:43 EST on 18 September 1964.

S-I-8

Summary

This was the first S-I booster to be made by the Chrysler Corporation and the first to be assembled at the new Michoud Operations. Launched from Cape Kennedy with the Pegasus B satellite on board.

Engines

S-I-8 and S-I-10 in final assembly at Michoud (10.1963) 6415020

The initial and static firing engine configuration was:

Position 101: H-5019
Position 102: H-5020
Position 103: H-5021
Position 104: H-5022 – subsequently removed from stage and replaced by H-5032
Position 105: H-2016
Position 106: H-2017 – subsequently removed from stage and replaced by H-2032
Position 107: H-2018
Position 108: H-2029 – subsequently removed from stage and replaced by H-2031

Spare engine: H-5032 - installed in place of H-5022
Spare engine: H-2032 - installed in place of H-2017
Spare engine: H-2031 - installed in place of H-2029

The final flown engine configuration was:

Position 101: H-5019
Position 102: H-5020

Recovered portions of the I-B seal retainer ring from engine H-2029 (6.1964)

Position 103: H-5021
Position 104: H-5032
Position 105: H-2016
Position 106: H-2032
Position 107: H-2018
Position 108: H-2031

Stage manufacturing

On 4 October 1962 Chrysler Corporation Space Division (CCSD) began fabricating components for the S-I-8 stage. The H-1 engines were shipped from Rocketdyne to Huntsville. MSFC completed mechanical receipt inspection and functional analysis of the S-I-8 engines in February 1963 and began modifying the engines to the SA-8 configuration. This modification was completed in March and the engines were placed in storage at MSFC to await shipment to CCSD at Michoud Operations. In April CCSD began assembling the first industry-produced S-I stage and on 10 May completed clustering of the propellant tanks. In April and May the H-1 engines were shipped from Huntsville to Michoud. The inboard engines were installed in late August 1963 and the outboard engines in September 1963.

In early October 1963 CCSD at Michoud completed final assembly of the S-I-8 booster stage. On 27 October CCSD transferred S-I-8 from the assembly area to the checkout area and began engine checkout and final checkout of the stage.

At a ceremony in Michoud on 13 December 1963 Lynn A Townsend, President of Chrysler Corporation, presented the first industry-built Saturn S-I stage to Dr Wernher von Braun, Director of MSFC.

At MAF, critical tubing assemblies were replaced dur-

The S-I-8 stage passing in front of the Administration building at Michoud (17.4.1964)

Test firing of a S-1 stage at MSFC (1964) 6413477

ing January and February 1964. The S-I-8 completed pre-static checkout on 18 February 1964. It remained inside the blast curtain area until 19 March, when it was removed for cleaning, painting and preparation for shipment.

Stage testing

On 17 April 1964 the S-I-8 stage was shipped from MAF to MSFC in Huntsville on board the barge Promise. It arrived at MSFC on 25 April and the stage was erected in the Static Test Tower East on 27 April. A propellant loading test was performed on 12 May 1964. All LOX pre-valves were exchanged for a type modified to relieve pressure build-up between pre-valves and pump. LOX was loaded on 21 May for the short duration run but LOX leaks at the center tank lower manhole cover caused postponement. The leaks were cor-

Handover ceremony for the S-I-8 stage at Michoud
(13.12.1963)

S-I-8 being loaded aboard the barge Promise at Michoud
(17.4.1964)

**The S-I-8 stage on the barge Promise
passing downtown New Orleans** (17.4.1964)

rected and another LOX loading test was successfully conducted on 25 May.

The first static firing, SA-20, was initiated at 1642:23.38 CDT on 26 May 1964 with a duration of 48.94 seconds (inboard engines). The cut-off signal was initiated by the firing panel operator. During the test a leak occurred in the calibration valve of fuel tank # 4 causing fuel to be sprayed onto cables and instruments below this fuel tank. The fuel tank subsequently underwent cleaning. Also, due to abnormally high thrust levels, six of the eight engines (numbers 102, 103, 104, 106, 107 and 108) were re-orificed after this test.

The full duration firing, SA-21, was initiated at 1639:55.418 CDT on 11 June 1964 with a duration of 145.61 seconds (outboard engines), 139.92 seconds (inboard engines). There was a loss of thrust on engine # 108 (H-2029) which occurred at 48 and 136 seconds of main-stage firing. This engine was removed from the stage for examination and replaced with a spare engine (H-2031). It was discovered that the screws that hold the I-B retainer in the turbine first stage cavity had come out causing damage to the turbine blades. Inspection of the turbine on engine H-5011 (not on this stage but used for single engine tests in the Power Plant Test Stand between 27 May and 3 June 1964) also revealed four loose screws on the I-B seal. This engine (H-5011) was eventually installed on the SA-D booster stage on display at MSFC. The deficiency led to a decision to remove and inspect all turbines from stages S-I-7, S-I-8 and S-I-9. Meanwhile the engine that had the original problem (H-2029) was never used again, became an R&D spare and was transferred to the Smithsonian.

One of the re-orificed engines, # 106 (H-2017), did not respond normally to the re-orificing as witnessed by a second out-of-specification thrust value. As a result the engine was removed from the stage. A spare engine (H-2032) was shipped from Michoud to Huntsville. The engine that was removed (H-2017) was subject to two

60 second engine-level calibration firings in the Power Plant Test Stand (P1-438 on 9 July 1964 and P1-439 on 10 July 1964) in order to find out why it had not responded to re-orificing. Engine H-2017 was never used again, and became an R&D spare before transfer to the Smithsonian.

The two replacement engines for the S-I-8 stage were calibration tested in the Power Plant Test Stand in single engine tests before being inserted in the stage. Engine H-2031 was fired for 60 seconds on 18 June 1964 (test P1-435) and engine H-2032 was fired for 60 seconds on 27 June 1964 to calibrate the engine (test P1-436) and again for 60 seconds on 30 June (test P1-437) to re-calibrate the engine after the turbine repair.

A LOX loading test on 17 June investigated the problems encountered when loading to an ullage of 2.2%. Test data indicated overflow in the GOX riser line during helium bubbling and procedure changes were implemented.

The stage was removed from the test stand on 23 June and departed Huntsville on board the barge Promise on 24 June, arriving at Michoud on 29 June 1964. CCSD began post static modification and repair in July. During visual inspection it was found that a large number of bolts in the four antenna panels attached to the LOX tanks had become sheared. Further inspection revealed that the same had occurred on S-IB-9. The cause of the shearing was determined to be the contraction of the LOX tanks. During August all the H-1 engines were returned to Rocketdyne's Neosho plant to be retrofitted with new LOX domes. The engines were returned and re-installed in the stage prior to the completion of modifications at the end of September.

The stage was moved into station # 1 for electrical checkout at MAF on 13 October 1964. An initiator metering valve from the # 108 engine was removed and replaced in order to secure proper test results during post static testing. Disassembly of the valve showed that it contained traces of water. As a result all other valves were removed and inspected.

The LOX dome retrofit modification was accomplished on all engines at this time. Engine # 102 (H-5020) was damaged to the extent that two dents in the thrust chamber had to be brazed.

Engine 104 (H-5022) was damaged when a pair of pliers fell and pierced two thrust chamber tubes. The engine was removed and returned to Rocketdyne for repairs. The engine was replaced on the stage by spare engine H-5032. Meanwhile the repaired engine (H-5022) was used as a spare for the S-I-7 and S-I-9 stages before being transferred to the Atlas program.

Because of delays to the SA-9 upper stage the shipment of the S-I-8 stage to KSC was delayed from 19 December 1964. Post static checkout was completed on 22 December 1964. Following this the stage was weighed and the center of gravity determined prior to the attachment of the fins.

On 11 February 1965 handling equipment dented a fuel tank. This was overcome by pressing the surface smooth. Later, workers found various scratches, dents and gouges on the GOX lines and the suction line bellows. The GOX line was replaced but the other observations were considered minor. The stage was shipped from Michoud to Cape Kennedy aboard the barge Promise on 22 February 1965, arriving on 28 February.

Prior to pad erection at the launch site workmen corrected such S-I-8 discrepancies as a damaged fuel pressurization switch, removal of tension ties, contamination of the hydraulic fluid, moisture problems with the helicon connectors and missing "O" rings. Contamination found in the H-1 engine gas generators resulted in the removal of the generators and their return to Michoud for cleaning before re-installation.

The S-I-8 stage was erected on LC-37B on 2 March 1965, followed by the S-IV-8 stage and S-IU-8 on 17 March. The Apollo BP-26 CM, SM and adapter arrived at KSC on 10 April. The Pegasus B satellite arrived on 15 April and the assembly was erected on the launch vehicle on 28 April. Cryogenic tanking was completed on 11 May, the Flight Readiness Test on 14 May and the CDDT on 20 May. The S-I-8 fuel was loaded on 18 May and the LOX on launch day.

The SA-8 vehicle, including the S-I-8 booster stage, lifted off at 0335:01 EDT on 25 May 1965.

S-I-9

Summary

Final S-I stage to be manufactured and assembled at MSFC in Huntsville. Launched from Cape Kennedy together with the Pegasus A satellite.

Engines

The engine configuration in the stage for static testing and for launch was:

Position 101: H-5023
Position 102: H-5012
Position 103: H-5025
Position 104: H-5026
Position 105: H-2020
Position 106: H-2022
Position 107: H-2023
Position 108: H-2024

Spare engine: H-5022 (not used, transferred to Atlas program)

Stage manufacturing

MSFC started S-I-9 component fabrication in Huntsville and continued through to May 1963. On 4 June 1963 MSFC began clustering the propellant tanks. Also in June MSFC reworked the inboard engines to conform to the S-I-9 configuration. They also received and inspected the outboard engines. The ME Division completed the tank clustering on 19 June 1963. The inboard engines were installed in the stage by 17 July and during August 1963 MSFC completed installation of the outboard engines in the S-I-9 stage in Huntsville.

One outboard engine, accidentally damaged on 14 August, was replaced by a spare. All engines were in place by 28 August. In September MSFC completed the rest of the stage assembly except for one flight component not yet delivered. As a temporary measure a nonflight component was installed in its place. Quality Laboratory began pre-static checkout of the S-I-9 stage on 1 October 1963. This was interrupted in December 1963 in order to replace critical tubing assemblies after the problem on SA-5.

At MSFC replacement of the critical tubing assemblies was completed on 27 January 1964. Pre-static checkout was resumed and completed on 11 February 1964.

Stage testing

The S-I-9 stage was installed in the Static Test Tower

Test firing of a S-I stage at MSFC (1964) 6414826

East at MSFC on 17 February 1964. A simulated flight sequence test was completed on 2 March 1964. The following day a propellant loading test was performed.

The first static firing, SA-18, was attempted on 12 March 1964. However it was aborted, without any firing, when the Conax valves were inadvertently fired. The stage firing was rescheduled for the following day, and initiated at 1634:50.49 CST on 13 March 1964. The test was partially successful with a firing duration of 35.22 seconds (inboard engines). However, performance was low on all engines except engine # 102. Only engine # 103 had a thrust that was actually outside the tolerance limits. The seven engines were subsequently re-orificed to up-rate their performance. The cut-off was initiated by the firing panel operator.

The full duration firing, SA-19, was performed at 1335:32.59 CST on 24 March 1964. The test duration was 142.21 seconds (inboard engines), with all engines performing at the nominal thrust level. The test firing was witnessed by Mrs. Lyndon B Johnson, the President's wife, as well as NASA Administrator James E Webb. Cut-off signal was initiated by the flight sequencer 2 seconds after closure of the LOX low level sensor in tank # 4.

The S-I-9 stage was removed from the MSFC test stand on 8 April 1964 and returned to inspection. On 5 May the stage was returned to Quality and Reliability Assurance Laboratory for post static checkout. The checkout was interrupted in July following the discovery of cracks in the LOX domes of the S-I-7 H-1 engines. All the S-I-9 engines were removed from the stage and shipped to Rocketdyne in Neosho for replacement of the LOX domes and for fuel shaft seal replacements. The engines were returned to MSFC in late August. Following reinstallation of the engines in the stage the post static checkout was resumed.

A one month delay in the launch schedule allowed for

time to install hardware in the S-I-9 stage that would normally have been attached at KSC.

On 19 October 1964 MSFC shipped the S-I-9 stage, the S-1 stage fins and the S-IU-9 to KSC on board the barge Promise. They arrived safely on 30 October and were transferred to Hangar AF. The S-I-9 stage was erected on LC-37B on 3 November followed by the S-IV-9 and S-IU-9 stages on 19 November. Electrical mating of the three stages took place on 14 December. With the Apollo CM BP-16 already at the Cape, the SM and its adapter arrived from MSFC via the Pregnant Guppy on 13 November. The Pegasus A satellite arrived at KSC on 29 December 1964 after leaving General Electric's Valley Forge, Pennsylvania plant. The Pegasus A was erected on top of the S-IV-9 stage on 13 January 1965 and a day later the Apollo spacecraft was installed.

The All Systems Test ended on 5 February 1965 and the CDDT took place on 12 February. The SA-9 vehicle, with the S-I-9 booster stage, lifted off from LC-37B at 0937:03 EST on 16 February 1965.

S-I-10

Summary

The second S-I stage to be assembled by the Chrysler Corporation in Michoud, and the final S-I booster stage to be built. Launched from Cape Kennedy, including the Pegasus C satellite.

Engines

The engine configuration in the stage for static testing and for launch was:

Position 101: H-5028
Position 102: H-5029
Position 103: H-5030
Position 104: H-5031
Position 105: H-2034
Position 106: H-2026
Position 107: H-2027
Position 108: H-2030

Stage manufacturing

On 4 October 1962 CCSD began fabricating components for the S-I-10 stage at Michoud. In January 1963 CCSD began fabricating the upper and lower thrust ring assemblies and the fin and thrust support outriggers. During April CCSD finished building the stage barrel assembly and the thrust and fin outriggers. Build-up of

**S-I-10 stage being loaded onto the
barge Promise at Michoud** (24.7.1964)

the S-I-10 tail section was completed on 25 June and assembly of the tail section started the following day.

On 25 July 1963 CCSD initiated the clustering of tanks for the S-I-10 stage. The fuel tanks were installed early in September. Meanwhile, Rocketdyne delivered the H-1 engines for this stage. During the last quarter of 1963 CCSD inspected and completed modifications to the engines and continued stage assembly. Critical tubing assemblies were replaced on the stage following earlier problems on SA-5. Assembly of the stage was completed at Michoud in March 1964. The S-I-10 stage went into checkout on 4 May 1964. A small fire in the stage was caused by an overheated drill. Stage checkout was completed on 13 July 1964. Because of difficulty in the modification of the pre-valves for this stage it was decided to ship the valves to MSFC separately from the stage.

Stage testing

The S-I-10 stage was shipped from Michoud to Huntsville on the barge Promise. It left Michoud on 24 July 1964 and arrived in Huntsville on 31 July. It was installed in the Static Test Tower East at MSFC on the same day. Shortly afterwards all the H-1 engines were removed from the stage and returned to Rocketdyne's Neosho plant to be retrofitted with new LOX domes. All the engines were returned to MSFC after acceptance firings at Neosho. The 8 H-1 engines were reinstalled in the S-I-10 stage between 24 and 29 August 1964.

Prior to the engine installation a special LOX loading test was performed on 19 August to determine LOX boil-off rates and effectiveness of helium bubbling at various rates and durations.

A second propellant loading and flight sequence test was performed on 9 September 1964. The main objec-

tives were to perform leak checks, determine pre-valve activation times at cryogenic temperatures, and further investigate LOX overflow with 2.2% ullage.

The first attempt to perform the 35 second duration firing, on 16 September 1964, was terminated due to a LOX leak at the center LOX tank rear cover flange. Firing SA-22 was performed at 1637:57.22 CDT on 22 September 1964, but was aborted after 3.01 seconds with cut-off being automatically initiated at "time for commit" because the "all engines running" relay was not energized. This was due to a lack of the thrust OK signal from engine # 101, caused by a backflow through the position # 101 gas generator LOX injector purge check valve. Inspection of the check valve revealed a small burr on the inner surface of the valve body and a slight galling on the poppet. It was concluded that the check valve did not seat in the closed position.

The valve was subsequently replaced and firing SA-23 was performed successfully at 1637:06.83 CDT on 24 September 1964 for a duration of 35.08 seconds. Inboard engine cut off was initiated by the firing panel operator. There were 66,517 gallons of LOX on board at the start of the test and 50,401 gallons at the end. There were 42,104 gallons of RP-1 on board at the start and 32,213 gallons at the end. After this firing, cracks in the test stand hold-down bracket lower clevis ears at positions 1 and 3 were discovered. Although it was determined that the cracked brackets did not need to be replaced it was decided to eliminate a portion of the gimbal program during the next test to avoid unnecessary strain on the brackets.

The long duration test, SA-24, was started at 1638:34.13 CDT on 6 October 1964 and lasted for 149.93 seconds (inboard engines), 154.48 seconds (outboard engines to LOX depletion). There were 64,560 gallons of LOX on board at the start and zero at the end. There were 42,104 gallons of RP-1 at the start and 1,880 gallons at the end.

During the static firing SA-24 engine # 106 showed combustion chamber pressure oscillations. The honeycomb heat shield panel instrumentation was severely damaged. After investigation the polyurethane foam enclosed in the torque boxes of S-I-10 were made fire resistant by the use of additives. Following completion of the acceptance tests CCSD conducted a series of propellant loading tests on the stage to investigate problems associated with loading LOX to 2.2% ullage. These tests ended in October.

Following static firing the S-I-10 stage was removed from the test stand on 29 October and loaded on the barge Palaemon on 2 November. It was shipped from MSFC to Michoud, arriving on 7 November 1964. The booster stage then underwent final checkout and preparations for shipment to KSC. However, a bag of tools was dropped on the thrust chamber of outboard engine # 103 (H-5030) creating a hole in the tubing. A calorimeter was being installed on the engine at the time of the accident. The engine was dismounted from the stage, and returned to MSFC in Huntsville for repair and re-certification. After repair the engine was installed in the Power Plant Test Stand and underwent a 60 second verification firing (P1-457) on 10 December 1964. The engine was returned to Michoud on 12 December and reinstalled in the S-I-10 stage. To prevent the problem from occurring again CCSD took steps to ensure that engine covers remained in place when work was being performed on the engines. In addition, mechanics working in the engine areas would have their tool bags attached to their bodies.

CCSD moved the S-I-10 stage into the Checkout Station # 1 at Michoud on 8 January 1965 and began post static checkout and repair of the stage. Repairs included replacement and verification of a LOX pump cavity seal in engine # 105 that had shown excessive leakage and the smoothing of wrinkles caused by buckling of fuel tank # 3. CCSD completed post static checkout and

repair on 9 March and stored the stage to await shipment to KSC.

The S-I-10 stage was shipped from Michoud by the barge Promise on 26 May, arriving at Cape Kennedy on 31 May 1965. The S-IV-10 stage arrived on 10 May and the S-IU-10 on 1 June. The S-I-10 stage was erected on LC-37B on 2 June, the S-IV-10 on 8 June and the S-IU-10 on the following day.

The Apollo BP-9A SM and SM adapter arrived from MSFC on 21 June The Pegasus C satellite arrived from FHC on 22 June and the BP-9A CM and LES on 29 June. The BP-9A had been converted from BP-9, which had been used in Saturn I dynamic testing. The Pegasus C, adapter and CSM were erected on 6 July and the LES on 8 July, completing the satellite assembly. Simulated LOX and LH2 loading tests were performed on 13 July, and the Flight Readiness Test took place seven days later. Flight loading of RP-1 fuel took place on 23 July and the CDDT was successfully performed on 26 and 27 July.

The SA-10 vehicle, including the S-I-10 booster lifted off from LC-37B at 0800:00 EDT on 30 July 1965.

Firing SA-24 of the S-I-10 stage at MSFC (6.10.1964) 6413599

S-I-111

Summary

Planned first operational Saturn I booster cancelled at an early stage and replaced by the Saturn IB.

Engines

No engines ever installed

Stage manufacturing

During July to September 1963 CCSD fabricated the thrust support outriggers, fin support outriggers and barrel section for this first operational S-I stage. In August CCSD began fabrication of the S-I tail section and the second stage adapter and continued through October. At this time CCSD received the 70-inch fuel tanks and the 105-inch oxidizer tank and planed for the clustering to start on 29 October. However, on 30 October 1963 NASA announced its decision to rephrase the manned flights and delete the Saturn I operational vehicles. Components were transferred to the S-IB-1 stage.

S-I-112

Summary

Planned second operational Saturn I booster cancelled at an early stage and replaced by the Saturn IB.

Engines

No engines ever installed

Stage manufacturing

Initial manufacture of components for the S-I-112 stage commenced in July 1963. However, on 30 October 1963 NASA announced its decision to rephrase the manned flights and delete the Saturn I operational vehicles. Components were transferred to the S-IB-2 stage.

S-I-D5 (S-IB-D/F) (part of SA-D5 onwards)

Summary

Dynamic Test stage used to qualify the Saturn I Block

II and Saturn IB vehicles. Tested in the Dynamic Test Stand at MSFC. Used for launch pad interface checks at Cape Canaveral. Currently on display at the USSRC in Huntsville.

Engines

Eight H-1 engines were installed.

Stage manufacturing

MSFC completed assembly of the booster stage for the block II SA-D5 vehicle, designated S-I-D5, on 29 October 1962. Work had started in July 1962. The booster stage was installed in the MSFC Dynamic Test Tower on 13 November 1962. On 26 November the S-IV-D5 was erected atop the booster and by 17 December MSFC had completed the assembly of the IU, payload adaptor and payload body on the ground. This configuration simulated the SA-5 launch vehicle.

Stage testing

Simulation of propellant mass distributions for the boost flight conditions were achieved by filling the booster tanks with de-ionized water on an equal weight basis. The booster was suspended vertically in the test stand by spring clusters attached to hydraulic lifters attached to the test tower that allowed the activation forces to be applied. Phase I dynamic testing was

Installation of the S-IB-D/F stage in the dynamic stand at MSFC (11.1.1965) 6517817

Dynamic testing of the SA-202D configuration dynamic stages in the stand at MSFC (14.4.1965) 6522399

Procession of stages, including the S-IB-D/F at rear, heads to the ASRC (28.6.1969)

delayed from 17 December 1962 until January 1963, to allow for completion of revisions to the test tower. Once the tower was completed, in January 1963, the IU, payload adapter and payload were mounted on top of the two stages already installed in the test tower.

The MSFC Test Division began the Phase I dynamics testing on 8 January 1963 and completed the test program on 8 March. The Phase I testing determined the bending modes in the pitch and yaw directions, torsional modes, resonance and frequency response.

On completion of the Phase I tests the S-I-D5 stage was hydrostatic pressure tested to verify its structural integrity. During the week of 18 March the SA-D5 vehicle was removed from the test stand. The S-I-D5 stage left Huntsville on 5 April 1963 on board the barge Promise, bound for Cape Canaveral, where it arrived on 15 April, for use in checkout of Launch Complex 37B. Phase II vibration testing, comprising the upper stages only, was performed at MSFC whilst the booster stage was at Cape Canaveral.

The booster stage was erected on LC-37B on 18 April, and the following day the S-IV Dynamics/Facilities vehicle was erected on top of the first stage.

The series of wet tests were designed to check out LC-

37B equipment involved in propellant loading operations. During the week of 24 April propulsion, calibration and mechanical checks were performed. In the first week of May the RP-1 loading tests, the fuel tank pressurization and liquid nitrogen and LOX line tests were completed. Checkout of the LC-37B ended during June. The S-I-D5 stage departed for Huntsville on board the Palaemon on 1 July,

Following the propellant loading tests at Cape Canaveral Launch Complex 37B the dynamic booster stage, redesignated S-I-D6, returned to MSFC on 14 July 1963. The upper stages (S-IV-D6, CSM (BP-9) and LES) of SA-D6 were removed from the MSFC dynamic test stand by 18 July 1963, so that the S-I-D6 stage could be installed. The booster's fins were installed in the test stand on 22 July. The S-I-D6 stage itself was installed in the test stand on 25 July. The S-IV-D6 stage was mounted atop the first stage on 30 July; the instrument unit was installed on 7 August and the CSM was mounted on 12 August. With the complete SA-D6 vehicle now installed in the test stand Phase III vibration testing was started on 21 August 1963. Testing was accomplished on 4 October and the vehicle was removed from the stand on 22 October 1963. It was determined that vibration testing in the SA-7 configuration would be unnecessary due to the similarity between SA-6 and SA-7.

Following the SA-D6 testing the MSFC Test Laboratory transferred the CSM and adapter to the ME Laboratory for modification to the SA-8 and SA-9 micrometeoroid capsule configuration. It was determined that only one series of dynamic tests would be necessary to cover the SA-8, SA-9 and SA-10 configurations due to their similarity.

Preparation of the Dynamic Test Stand for the SA-D9 complete vehicle tests was started on 23 April 1964. This followed SA-D9 upper stage dynamic testing between 13 March and 4 April 1964. The S-I-D9 stage was installed on 29 April and the S-IV-D9 stage was installed on 6 May. Dynamic testing began on 20 May

Assembly of the S-IB-D/F stage at ASRC (6.1969)

and ran through to 9 July.

Disassembly of the vehicle began on 10 July and was completed by 17 July. The S-I-D9 stage was shipped from MSFC to Michoud, for modification, updating and checkout, arriving on 22 July 1964. This activity was performed in station # 3 at MAF. It was to be converted into the S-IB-D/F configuration. CCSD weighed the stage to determine the longitudinal center of gravity and then removed the components and tanks. One fuel tank and the 105 inch LOX tank were shipped to Ling-Temco-Vought in Dallas, Texas, for modification. The other components were labeled S-IB-D/F and stored.

In late September 1964 CCSD began build up of the S-IB-D/F stage with a modified flight tail section, a new spiderbeam, the modified tanks and the useable S-I-D9 components. Ballast was added in the stage to relocate the centre of gravity to the S-IB-1 configuration. CCSD completed the modification and checkout of the stage in early December and the stage was shipped from MAF to MSFC on 22 December 1964 on board the barge Palaemon in preparation for dynamic testing. The Palaemon was in a combined tug tow together with the Promise which had the S-IVB-D stage on board. It arrived at MSFC on 4 January 1965.

The S-IB-D/F Dynamic Test stage underwent Saturn IB vehicle testing at MSFC during the first five months of 1965. The Saturn IB flight vehicle had four flight configurations that differed enough for each one to require a separate series of dynamic tests.

SA-201, SA-202, SA-204, SA-205 configuration consisted of the launch vehicle plus the Apollo spacecraft without the LM
SA-203 configuration had no spacecraft and consisted of the launch vehicle and a simple nose shroud
SA-206 configuration incorporated a LM and a boilerplate CSM (BP-27)
SA-207 configuration consisted of the launch vehicle

S-IB-D/F stage at USSRC (2008)

and a complete Block II spacecraft

Dynamic testing was performed in the modified Saturn I Dynamic Test Stand at MSFC. The first stage of the vehicle comprised the Saturn I dynamic test stage (SA-D5) modified to the Saturn IB configuration (S-IB-D/F). The S-IVB-D stage was the upper stage of the vehicle.

Both the S-IB-D/F and the S-IVB-D stages were installed in the Dynamic Test Stand in January 1965. On 8 February the dynamic test instrument unit, S-IU-200D/500D was installed atop the S-IVB-D stage. The SLA (simulating the nosecone) was airlifted in by helicopter from NAA's Tulsa plant and placed on top of the

S-IB-D/F stage engines at USSRC (2008)

dynamic test vehicle, although this operation was delayed due to damage sustained by the SLA on landing.

The first phase of testing (SA-203 configuration) started on 18 February 1965 and lasted until 2 March. The second phase of testing (SA-202 configuration) included Boilerplate CSM, BP-27, and started on 15 March. On 27 March the S-IB-D/F spider-beam assembly crossbeam web cracked. This failure necessitated repair of the spider-beam and repeat of some of the SA-202 tests. Dynamic testing resumed on 2 April and continued until 19 April. The third phase of testing (covering the SA-207 configuration) ran from the end of April until 12 May. The fourth and final phase of dynamic testing (SA-206 configuration) ended on 27 May 1965, concluding the Saturn IB total vehicle test program. Because of the damage to the spiderbeam it was decided not to send the S-IB-D/F stage to Cape Kennedy (originally planned for 15 May) for pad facility verification. Instead the S-IB-1 stage was used for that task. Vehicle disassembly began immediately in preparation for upper stage tests. Meanwhile, the S-IB-D/F stage was placed in storage at MSFC.

Sometime between May 1965 and September 1966 the S-IB-D/F stage was shipped from MSFC to MAF for storage.

On 23 September 1966 the S-IB-D/F stage was shipped from MAF back to MSFC on board the barge Promise. It had been in storage at MAF for some time. Once back at MSFC it was placed on display at the MSFC Space Orientation Center.

Ground breaking for the Alabama Space and Rocket Center in Huntsville began in July 1968. The following year the S-IB-D/F stage was moved to the outdoor exhibit area. On 26 June 1969 it was moved to a position near the Astronautics Lab at MSFC. Two days later, on Saturday 28 June at 0500 CDT, the stage was transported along Rideout Road to the museum. A number of power lines had to be disconnected, road signs taken down and some poles moved. The stage was mounted in a vertical orientation with an S-IV stage (Hydrostatic/Dynamics vehicle) and payload (BP-27) on top. The museum opened its doors the following year and the stage has been in the museum ever since.

S-IB-1

Summary

First stage to use the up-rated 200,000 pounds thrust H-1 engines. First S-IB stage, used to launch the AS-201 mission which was a sub-orbital re-entry test. Initially

used for launch pad verification testing. One fuel tank had to be replaced on the launch pad because of over pressurization.

Engines

The initial and static firing engine configuration was:

Position 101: H-7046
Position 102: H-7047
Position 103: H-7048
Position 104: H-7049
Position 105: H-4044
Position 106: H-4045
Position 107: H-4046 - subsequently removed from stage and replaced by H-4052
Position 108: H-4047

Spare engine: H-4052 – installed in place of H-4046

The final flown engine configuration was:

Position 101: H-7046
Position 102: H-7047
Position 103: H-7048
Position 104: H-7049
Position 105: H-4044
Position 106: H-4045
Position 107: H-4052
Position 108: H-4047

S-IB-1 stage being hoisted into the MSFC test stand
(15.3.1965) 6520960

S-IB-1 stage in the MSFC test stand (15.3.1965) 6520963

S-IB-1 long duration firing at MSFC (13.4.1965) 6522248

The outboard engines were re-installed in the stage between 18 and 10 February 1965. The inboard engines were re-installed in the stage between 12 and 15 February 1965. All engines were electrically checked. On 6 March the stage was shipped from MAF to MSFC aboard the barge Promise. It arrived in Huntsville on 14 March and was immediately off-loaded and one day later installed in the Static Test Tower East at MSFC.

Stage testing

Preparations for static tests included two propellant loading tests on 24 and 29 March. These tests included leak checks and investigations of the LOX boil-off rates. While performing an engine purge checkout on engine # 103 (H-7048) it was discovered that the hose assemblies from the stage purge panel to the engine customer connect panel were improperly connected such that the fuel injector purge was connected to the gas generator LOX injector purge, the gas generator LOX injector purge to the LOX dome purge, and the LOX dome purge to the fuel injector purge. The hose assemblies were changed to the proper configuration and no contamination was detected. During the turbine exhaust system leak check two leaks were discovered at

Stage manufacturing

CCSD fabrication of components for the first Saturn IB booster stage, S-IB-1, other than those already available from the S-I-111 stage, began at MAF in March 1964. The tail section was completed on 12 April 1964. CCSD completed the S-IB-1 thrust structure and the second stage adapter in May.

Clustering of the S-IB-1 tanks was started on 18 June 1964, and the spider beam or second stage adapter was attached to the tanks on 22 June. The S-IB-1 stage went into checkout without some long lead items in order to avoid slippage of schedule.

On 24 July CCSD completed clustering of tanks for the S-IB-1 stage. All the up-rated (200,000 pounds thrust) H-1 engines were received at Michoud from Rocketdyne in late July and installed in the stage. However, in early November all the H-1 engines were returned to Rocketdyne's Neosho plant to be retrofitted with new LOX domes and injector modifications and to be re-fired.

The booster completed assembly without the engines on 20 November and moved into checkout station # 2 for pre-static checkout on 24 November 1964. Checkout operations were completed on 2 February 1965.

Distorted bellows in turbine exhaust duct, heat exchanger inlet engine H-4047 (4.1965)

engine # 106 (H-4045). Corrections were made to stop the leaks.

Following the initial propellant loading test the screens at the LOX and fuel pump inlets were removed. A piece of O-ring approximately 4.5 inches long was found in the fuel pump inlet screen on engine # 105 (H-4044). The O-ring came from a damaged facility fill and drain valve seal.

The S-IB-1 stage was successfully static fired two times. Test SA-25 took place at 1702:15.460 CDT on 1 April 1965 and lasted 35.174 seconds (inboard engines), 35.294 seconds (outboard engines), with cut-off initiated by the firing panel operator.

Test SA-26 took place at 1638:12.778 CDT on 13 April 1965 and lasted 138.210 seconds (inboard engines), 145.010 seconds (outboard engines). The inboard engine cut-off signal was initiated by the switch selector 2 seconds after the LOX low level sensor in tank # O-2 was uncovered. The outboard engine cut-off signal, triggered by dropout of the Thrust OK pressure switch (TOPS) on engine # 101 (H-7046), occurred 6.80 seconds later. Engine # 103 (H-7048) solid propellant gas generator ignition characteristics were abnormal for this test in that combustion was established slower than usual. The effect of the slight delay in combustion was considered negligible.

An immediate post test inspection of engine # 107 (H-4046) revealed an internal thrust chamber tube leak 13 inches below the injector. A closer inspection indicated a longitudinal crack 0.375 inch long by 0.063 inch wide. Consequently engine # 107 (H-4046) was removed from the stage on 27 April 1965 and replaced with the allocated spare engine (H-4052) on 3 May, during post static checkout at MAF. Engine H-4046 was later reworked and flew on the S-IB-2 stage. Visual inspection of engine # 108 (H-4047), after test SA-26, revealed a distorted bellows section in the turbine exhaust duct adjacent to the heat exchanger. The bellows section was replaced during the post static checkout of the stage at MAF.

The stage was removed from the stand on 19 April. During removal five ripples in fuel tank F-3 were observed. The ripples were approximately 12 inches long and ¼ inch deep. Repairs were undertaken. The stage was loaded aboard the barge Palaemon on 20 April for its return to MAF, where it arrived on 24 April.

On 11 June CCSD placed the stage in the check out area and began post static checkout and modification. The only major problem encountered was leakage of the turbo-pump LOX shaft seals on engines # 101 (H-7046) and # 102 (H-7047). Rocketdyne engineers replaced the seals at Michoud.

The Simulated Flight Test was performed between 12 and 14 July. Because of damage to the S-IB-D/F stage it was decided to substitute the S-IB-1 stage for launch pad facility verification tests. Consequently the post static checkout of the S-IB-1 stage was expedited and was completed on 20 July 1965. The stage was shipped to Cape Kennedy, together with the S-IU-200F/500F, aboard the barge Promise on 9 August, where it arrived on 14 August. It was the first space cargo to barge through the newly constructed Port Canaveral locks.

After completing pre-flight checkout of the S-IB-1 stage in Hangar AF it was erected on LC-34 on 18 August. The S-IB-1 served as a spacer for the S-IVB-200F/500F during propellant tankings to verify the LOX and LH2 loading systems. The S-IU-200F/500F was erected atop the S-IVB-200F/500F.

Vehicle checkout of LC-34 began on 18 August and, except for several days lost to Hurricane Betsy, progressed extremely well. During instrument compartment leak checks on 10 September over pressurization of the forward bulkhead on fuel tank # 8 of S-IB-1 occurred, requiring replacement of the tank on the launch pad on 29 September, using a tank from the S-IB-6 stage. Technicians completed the automatic computer controlled propellant loading of the S-IVB-200F/500F on 23 September 1965. On completion of the tests on 29 September, KSC technicians began dismantling the upper stages from the pad, leaving the S-IB-1 in place. The flight upper stage, S-IVB-201, was erected on top of the S-IB-1 stage on 1 October 1965.

The S-IU-201 was erected on top of the vehicle on 25 October, power was applied to the S-IB-1 stage on 21 October and electrical mating of the entire vehicle was achieved on 10 November. The first successful automatic LOX loading of the S-IB-1 stage during pre-launch checkout occurred on 30 November. Erection of the SLA-3 and CSM-009 took place on 26 December 1965. The Flight Readiness Review was completed on 21 January 1966 and the LES was attached three days later. The dry CDDT took place on 8 February and the wet CDDT the following day. The Flight Readiness Test was completed on 12 February and RP-1 tanking operations started on 19 February.

The AS-201 vehicle with the S-IB-1 first stage was launched successfully from launch complex 34 at 1112:01 EST on 26 February 1966.

S-IB-2

Summary

Second S-IB booster stage, but the third to be launched. Used to launch the AS-202 mission which was a sub-orbital re-entry test.

Engines

The initial engine configuration was:

Position 101: H-7051
Position 102: H-7052
Position 103: H-7053 – removed from the stage and replaced by H-7050
Position 104: H-7054
Position 105: H-4048
Position 106: H-4049
Position 107: H-4050
Position 108: H-4051

Spare engine: H-7050 – installed in place of H-7053

The static firing engine configuration was:

Position 101: H-7051
Position 102: H-7052
Position 103: H-7050
Position 104: H-7054
Position 105: H-4048
Position 106: H-4049
Position 107: H-4050
Position 108: H-4051 – subsequently removed from stage and replaced by H-4046

Spare engine: H-4046 – installed in place of H-4051

The final flown engine configuration was:

Position 101: H-7051
Position 102: H-7052
Position 103: H-7050
Position 104: H-7054
Position 105: H-4048
Position 106: H-4049
Position 107: H-4050
Position 108: H-4046

Stage manufacturing

CCSD fabrication of components for the second Saturn IB booster stage, S-IB-2, other than those already available from the S-I-112 stage, began at MAF in March 1964. CCSD accomplished build up of the spider-beam for the S-IB-2 stage in July and August 1964. In late September CCSD began the S-IB-2 assembly opera-

Erection of the S-IB-2 stage at KSC (4.3.1966)

tions with tank clustering.

CCSD completed assembly of the second S-IB flight booster on 26 February and moved it from the assembly area at Michoud to Checkout Station # 2. In January all eight H-1 engines were removed from the stage and returned to Rocketdyne's Neosho plant for LOX dome and injector retro-fit. During inspection in the week of 11 January, following the removal of the engines from the stage, it was discovered that four inboard engine thrust chambers were damaged. Re-installation of the repaired engines followed completion of the checkout on 27 April, and preparations commenced for shipping the stage to MSFC for static firing.

However, on 21 May 1965 a large dent was discovered on the aspirator of engine # 103 (H-7053) resulting in the return of that engine to Neosho for repair. The damage had been caused by impact with a Lift-a-Loft. The engine was replaced by a spare, H-7050, which was installed in the stage on 1 June 1965. Meanwhile, the damaged engine (H-7053), once repaired, became a FWV spare and was eventually scrapped. The S-IB-2 stage was shipped from MAF to MSFC on 12 June, arriving on 19 June.

Stage testing

During preparations for installation of the stage in the test stand several large dents appeared in the black painted section of fuel tank # 2. The dents appeared when the stage was exposed to sunlight and disappeared when the tank was shaded. No corrective action was implemented. The S-IB-2 stage was installed in the Static Test Tower East at MSFC on 21 June and a propellant loading test was performed successfully on 30 June.

The first attempt to static fire the stage, SA-27, occurred at 1641:24.827 CDT on 8 July 1965 but ended after only 3.002 seconds (inboard engines), 3.123 seconds (outboard engines), of the planned 35 seconds. This was due to a malfunction of the thrust OK pressure switch # 2 on engine # 104 (H-7054), which caused an automatic thrust failure cut-off.

The second attempt, SA-28, at 1636:23.820 CDT on 9 July 1965 was successful with a firing lasting 35.192 seconds (inboard engines), 35.302 seconds (outboard engines). The cut-off was initiated by the firing panel operator. At the start of this test there were 66,940 gallons of LOX and 42,100 gallons of fuel on board the vehicle.

The final, long duration firing, SA-29, took place at 1435:59.166 CDT on 20 July 1965 and lasted 143.285 seconds (inboard engines), 144.282 seconds (outboard engines). Inboard engine cut-off was given after the LOX low level sensor in LOX tank O-2 was uncovered. The outboard engine cut-off was initiated by the switch selector approximately one second after the inboard engine cut-off. At the start of this test there were 66,940 gallons of LOX and 42,100 gallons of fuel on board the vehicle. At the end of test this had decreased to 3,165 and 2,065 gallons respectively.

All engines produced thrust within the specified range of 200,000 pounds +/- 3% and no re-orificing was necessary. However, following the final, long duration firing, a total of five permanent ripples were found in fuel tank F-3. These were attributed to differential thermal stresses caused by exposure to the exhaust plume. The modified gimbal boots which utilized a layer of aluminized fabric bonded to the inner boot were burned severely during the static test. Severe metal erosion on the lower edge of the aspirator lip of engines # 103 (H-7050) and 104 (H-7054) was detected. It was decided not to make any repairs for this engine erosion.

The stage was removed from the test stand on 29 July and it left MSFC on board the barge Palaemon the following day, arriving at MAF on 7 August. At Michoud it underwent post static modification and repair. The stage was placed in Checkout Station # 2 on 1 October for post static checkout.

Post static checkout revealed that leakage of gear case oil through the LOX pump cavity "T" valve and check valve through the final static firing had contaminated inboard engine # 108 (H-4051). Technicians replaced this engine with a spare (H-4046), which had previously been installed in the S-IB-1 stage during static testing. The removed engine (H-4051) was eventually scrapped. During October and November technicians completed removal, cleaning and re-installation of 65 tube assemblies and fittings which had also been contaminated.

On 1 February 1966 the S-IB-2 stage departed MAF

bound for Cape Kennedy aboard the barge Promise. It arrived at KSC on 7 February 1966. The S-IB-2 stage was erected on LC-34 on 4 March. The S-IVB-202 stage, which had arrived at KSC on 31 January, was erected on top of the booster stage on 10 March. S-IU-202 was erected the following day. The SLA-4 and CSM-011 were erected on 2 July.

The Flight Readiness Review took place on 11 August. The S-IB-2 stage formed the first stage of the AS-202 vehicle that was launched from complex 34 at 1315:32 EDT on 25 August 1966.

S-IB-3

Summary

This stage used to launch the AS-203 vehicle which was primarily used to check the movement of liquid hydrogen in zero-g in the upper stage.

Engines

The initial, static firing and final flown engine configuration was:

Position 101: H-7056
Position 102: H-7058
Position 103: H-7059
Position 104: H-7060
Position 105: H-4053
Position 106: H-4054
Position 107: H-4056
Position 108: H-4057

Stage manufacturing

The S-IB-3 stage was the first S-IB stage to incorporate all the planned modifications completing the stage

Assembling three of the first S-IB booster stages at Michoud (21.6.1965) 6522687

redesign from the S-I configuration. The new changes in the S-IB-3 stage included lightweight propellant tanks, titanium fuel pressurizing tanks, low differential pressure LOX venting system, honeycomb heat shield panels and reoriented inboard engine turbine exhausts.

Fabrication of S-IB-3 components began at Michoud in July 1964. By the end of the year CCSD had completed the S-IB-3 barrel assembly, and assembly of the outriggers to the barrel assembly was underway.

CCSD began clustering of the propellant tanks for the S-IB-3 stage at Michoud on 21 January 1965. The H-1 engines arrived at MAF ahead of schedule, but parts shortages delayed other stage installations. Installation of the H-1 engines in the stage was completed on 21 April and structural build-up of the stage was completed in early June. On 17 June the stage was moved into Checkout Station # 1 for the start of pre-static checkout.

Several problems occurred during checkout. Firstly, excessive LOX turbo-pump shaft seal leakage on engine # 102 (H-7058) resulted in replacement of the seal. Secondly, disconnection of a leaking tube assembly disclosed some metal chips and further analysis of the complete purge system revealed additional contamination by metal chips. This required purging the basket with missile-grade air.

Checkout operations were completed on 14 August, and after completion of modifications and shipping preparations, the stage was transported aboard the barge Palaemon, leaving Michoud on 9 September 1965 and arriving at MSFC on 16 September. Hurricane Betsy struck the area on the night of 9 September and the Palaemon with the S-IB-3 stage on board weathered the storm near Baton Rouge.

Stage testing

The S-IB-3 stage was in installed in the Static Test Tower East on 17 September and preparations for checkout and acceptance firing started immediately. The stage completed simulated flight tests on 29 September, 8, 11 and 21 October and a propellant loading test on 1 October.

Static firing SA-30 was performed at 1643 CDT on 12 October 1965 with a duration of 35.295 seconds. The full duration firing, SA-31, was performed at 1641 CDT on 26 October 1965 with a duration of 146.226 seconds. Cut-off occurred with LOX depletion as planned. All the tests were performed successfully and all the test objectives were met.

The stage was removed from the Static Test Tower East on 3 November and loaded on board a barge the following day for its return trip to MAF. It arrived at MAF on

9 November 1965 and entered post-static modification and repair.

The stage left MAF by the barge Promise on 7 April 1966 bound for Cape Kennedy. Upon arrival on 12 April it was unloaded and moved into Hangar AF later on the same day. On 15 April three horizontal fins were attached. On 18 April it was erected on LC-37B and on the following day the fin installation was completed. The S-IVB-203 stage was erected on 21 April followed by the S-IU-203 and nose cone on the same day.

Initial power was applied to the S-IB-3 stage on 25 April and a full pressure test was conducted on 26 May. On 7 June the fuel and oxidizer loading test was accomplished. The Flight Readiness Test was performed on 27 June and RP-1 fuel tanking operations started on the following day. The CDDT was performed between 29 June and 1 July.

The S-IB-3 stage formed the first stage of the AS-203 vehicle that was launched from complex 37B at 0953:17 EST on 5 July 1966.

S-IB-4

Summary

Originally planned that this stage launch Apollo1 from Pad 34 at Cape Kennedy. Following the Apollo 1 fire the stage was transferred to Pad 37B and launched the first lunar module, unmanned.

Engines

The initial and static firing engine configuration was:

Position 101: H-7062
Position 102: H-7063
Position 103: H-7064
Position 104: H-7065
Position 105: H-4058
Position 106: H-4059 – subsequently removed from stage and replaced by H-4062
Position 107: H-4060
Position 108: H-4061

Spare engine: H-4062 – installed in place of H-4059

The final flown engine configuration was:

Position 101: H-7062
Position 102: H-7063
Position 103: H-7064
Position 104: H-7065
Position 105: H-4058

Erection of the S-IB-4 stage on pad 37B at KSC (7.4.1967)

Position 106: H-4062
Position 107: H-4060
Position 108: H-4061

Stage manufacturing

CCSD commenced fabrication of S-IB-4 components on 18 October 1964. On 6 April 1965 CCSD positioned the S-IB-4 tail unit assembly into the main assembly cluster fixture at MAF. The 105 inch LOX container was attached to the tail unit on 8 April and on 13 April the spider-beam assembly was attached. Between 14 May and 9 August clustering of the remaining propellant tanks was accomplished. The inboard engines were installed in mid-July and the outboard engines were in place by 3 August. Buildup of the stage was completed by Chrysler on 6 October. Pre-static checkout was completed on 9 November prior to pre-shipment preparations.

The S-IB-4 stage was shipped from Michoud aboard the barge Palaemon on 7 December 1965, arriving at Huntsville on 13 December 1965. The following day the stage was unloaded from the barge and installed in the Static Test Tower East at MSFC.

Stage testing

After stage alignment and connecting the stage to ground equipment, functional tests and LOX and fuel system leak checks were performed. The hydraulic system of engine # 101 (H-7062) was found to be contaminated with rubber and metallic particles, probably orig-

S-IB-4 engines at KSC (7.4.1967)

inating from the accumulator. The complete system was removed and replaced by a spare system. Simulated flight tests with flight pressures were performed on 3, 4 and 13 January 1966. In addition a propellant loading test was performed on 5 January. The first static test was originally planned for 14 January 1966, but was cancelled because the yaw actuator on engine # 103 (H-7064) was not controllable from the blockhouse. A new yaw actuator was installed.

The stage underwent the first short static firing (SA-32) at 1644:10.636 CST on 17 January 1966 for a duration of 35.227 seconds (inboard engines), 35.339 seconds (outboard engines). Cut-off was initiated by the firing panel operator.

The second, long duration firing, SA-33, took place at 1640:19.820 CST on 21 January 1966 with a duration of 143.934 seconds (inboard engines), 147.110 seconds (outboard engines). Inboard engine cut-off was initiated by the switch selector 3.19 seconds after the LOX low level sensor # 3 in LOX tank O-4 was uncovered. Outboard engine cut-off was initiated by LOX depletion of engine # 103 (H-7064). All engines produced thrust within the limits of 200,000 pounds +/- 3% and no engine re-orificing was required. Ignition transient data was not recorded because the recorder room personnel did not hear the countdown times after X-30 seconds because of a defective headset being used by the person calling the countdown times.

The S-IB-4 stage was removed from the static tower on 28 January and loaded on the barge later the same day. On 29 January 1966 the stage left MSFC aboard the barge Palaemon, bound for Michoud for Post Static Checkout.

Following completion of checkout, on 10 August 1966 the stage left MAF on board the barge Promise bound for Cape Kennedy, where it arrived on 15 August. It was transferred to Hangar AF for receiving inspection.

The stage was erected on LC-34 in August 1966 followed by the upper stage, S-IVB-204, on 31 August 1966 as part of the planned first manned Apollo flight in February 1967. The CSM-012 was mounted on top of the stages. During this time on the pad, in December 1966, engine # 106 (H-4059) had to be removed and was replaced by the spare (H-4062). The replaced engine (H-4059) was eventually transferred to the RS-27 program. The investigation of H-1 engine turbine blades disclosed that the first stage turbine blades of engine H-4059 were produced by a vendor that had not been qualified.

The mission was cancelled when the capsule caught fire on 27 January 1967 killing the three astronauts on board. After the Apollo 1 fire the S-IB-4 stage was de-erected in 6 April 1967. The CM-012 was placed in permanent storage, whilst the SM-012 was eventually scrapped in May 1977 at Rockwell in Los Angeles. The first stage was re-erected, this time on Launch Complex 37B, on 7 April 1967 and underwent a requalification test program. The S-IVB-204 stage was erected on top on 10 April, followed by S-IU-204 on 11 April and SLA-7 which enclosed the first lunar module, LM-1, on 19 November 1967.

The Flight Readiness Test was performed on 23 December 1967 and the RP-1 fuel was loaded into the S-IB-4 stage on 14 January 1968. The CDDT was completed on 20 January 1968, after being terminated the day before.

The S-IB-4 stage was eventually launched at 1748:08 EST on 22 January 1968 when it became the booster stage of the Apollo 5, AS-204/LM-1 vehicle that launched the first lunar module.

S-IB-5

Summary

Stage used to launch the first manned Apollo mission, Apollo 7.

Engines

The initial, static firing and final flown engine configuration was:

Position 101: H-7066
Position 102: H-7067
Position 103: H-7068
Position 104: H-7069
Position 105: H-4063
Position 106: H-4064
Position 107: H-4065

Position 108: H-4066

FWV engine: H-7053

Stage manufacturing

Fabrication of the S-IB-5 subassemblies began on 8 January 1965. The tail unit assembly was positioned in the main assembly cluster fixture at MAF on 15 July 1965. On 5 August the 105 inch LOX container was attached to the tail unit. The spider-beam was attached to this container on 10 August and on 18 August clustering of the remaining propellant tanks was started. Clustering operations ended on 30 August. The inboard engines were installed on 15 October and the outboard engines on 27 October.

Stage buildup ended on 30 November and workmen moved the S-IB-5 stage to the checkout station for pre-static checkout and shipping preparations.

The stage was shipped from MAF aboard the barge Palaemon, arriving at MSFC on 27 February 1966. It was installed in the Static Test Tower East on the following day.

Stage testing

The stage underwent the first static firing (SA-34) at 1749 CDT on 23 March 1966 for 35 seconds. The stage underwent the long duration firing (SA-35) of 144.6 seconds at 1641 CDT on 31 March 1966. LOX depletion terminated this successful test.

On 10 August 1966 the stage was accepted by NASA but moved to in-house storage at MAF at the direction of NASA until such time that the stage was needed at KSC. Plastic covering was wrapped around the stage to protect the black painted areas from direct sunlight. On 31 August 1967 modifications and refurbishment start-

The S-IB-5 stage entering the Point Barrow at Michoud prior to the trip to KSC (25.3.1968) 6864857

ed. All the engines were removed and the turbo-pumps inspected for possible lube oil seal leakage. None was found and the engines were re-installed. Also at this time representatives of Rocketdyne and Union Carbide performed an inspection of each H-1 rocket engine assembly to verify that the correct turbine wheel materials had been used. Inspection revealed that the proper Haynes Stellite 21 material had been used in the first stage turbine wheels of all the S-IB-5 engines.

Post-storage re-verification checkout of the stage was started at MAF on 28 December 1967 with the Simulated Flight Test being performed on 4 March 1968. Preparation for shipment of the stage started on 13 March 1968. The stage was shipped to Cape Kennedy on 25 March 1968 on board the AKD Point Barrow, arriving on 28 March 1968.

Preparation of the S-IB propellant tanks at Michoud (1967)

The S-IB-5 stage was erected on LC-34 in April, followed by the S-IVB-205 stage which was erected on top of the S-IB-5 stage on 16 April. Following this the S-IU-205, SLA-5 and CSM-101 were stacked. The CDDT was conducted between 12 and 16 September. The Flight Readiness Test was conducted on 26 and 27 September and RP-1 fuel loading was started on 4 October.

The S-IB-5 stage was eventually launched at 1102:45 EDT on 11 October 1968 when it became the first stage of the AS-205 vehicle that launched the first manned Apollo spacecraft, Apollo 7.

S-IB-6

Summary

Stage originally manufactured and tested in 1966. This was the first S-IB stage with up-rated 205,000 pounds thrust H-1 engines. It was also the first stage to have all white propellant tanks to overcome the deformations seen on earlier black-painted tank sections. Prepared for launch at KSC at the time of the Apollo 1 fire in 1967 before entering 5 years' storage. Finally launched in 1973 as the first stage of the first manned Skylab vehicle.

Engines

The initial and static firing engine configuration was:

Position 101: H-7071
Position 102: H-7072
Position 103: H-7073
Position 104: H-7075
Position 105: H-4068
Position 106: H-4069
Position 107: H-4070

Position 108: H-4071 – removed from S-IB-6 and installed in S-IB-8

Spare engine: H-4072 – utilized in place of H-4071

The final flown engine configuration was:

Position 101: H-7071
Position 102: H-7072
Position 103: H-7073
Position 104: H-7075
Position 105: H-4068
Position 106: H-4069
Position 107: H-4070
Position 108: H-4072

Stage manufacturing

CCSD personnel began clustering propellant tanks for the S-IB-6 stage on 25 October 1965. One of the fuel tanks from the S-IB-6 vehicle had been transferred to the S-IB-1 stage following an accident and damage to that stage at the launch site in September 1965. The first engine, # 105 (H-4068), was installed in the stage on 31

Initial S-IB stage assembly at Michoud (1967)

Clustering of the S-IB-6 propellant tanks at Michoud
(10.1965)

Inboard engine installation in the S-IB-6 stage at Michoud
(31.12.1965)

December 1965. The remaining engines were installed during the next four weeks with the final one being attached on 28 January 1966.

Stage testing

Post manufacturing checkout of the S-IB-6 stage was completed on 3 May 1966 and the stage was shipped from Michoud on 19 May, arriving at MSFC on 28 May. The stage was installed in the Static Test Tower East on 31 May. Simulated flight tests with flight pressure were performed on 15 and 21 June. On 16 June a propellant loading test on the stage was accomplished successfully. Prior to the first firing all the turbo-pump LOX seal assemblies were replaced with modified seal assemblies.

The first static firing (SA-36) took place at 1639:57.000 CDT on 23 June 1966 and lasted 35.464 seconds (inboard engines), 35.580 seconds (outboard engines). Cut-off was initiated as planned by the firing panel operator. Engine operation was satisfactory with all engines producing a thrust within the specified limits of 205,000 +/- 3,000 pounds and no re-orificing was required.

The second firing, (SA-37) for a full duration of 138.580 seconds (inboard engines), 141.236 seconds (outboard engines), took place at 1640:00.000 CDT on 29 June 1966. The cumulative total firing times for the outboard engines was 176.8 seconds, and 174.0 seconds for the inboard engines. Inboard engine cut-off was initiated by the switch selector, 3.176 seconds after the LOX low level sensor # 2 in LOX tank O-2 was uncovered. Outboard engine cut-off was initiated by dropout of an engine TOP switch as a result of LOX depletion. Engine operation was satisfactory, although all engine thrust values were below those predicted from the short duration firing. The thrust values for engines # 102 (H-7072) and # 104 (H-7075) were below specified limits.

Both firings took place in the Static Test Tower East at MSFC and were largely successful. However, there were problems with engine # 108, H-4071, during the second firing which resulted in that engine being removed from the stage just after the second firing on 29 June. The engine had a split thrust chamber tube in

Clustering of the S-IB-6 propellant tanks at Michoud
(10.1965)

**Completion of inboard engine installation
on the S-IB-6 stage at Michoud** (1.1966)

**Final engine being installed
in the S-IB-6 stage at Michoud** (28.1.1966)

S-IB-6 stage checkout at Michoud (1966)

an area above the throat and it was determined that the damage was not field-repairable. The following day a replacement engine, H-4072, was dispatched from MAF and installed in the stage (in the test tower) as soon as it arrived at MSFC on 1 July 1966. Thus, the S-IB-6 stage was only ever test fired with 7 of the 8 engines it finally flew with. The removed engine, H-4071, was returned to Rocketdyne's Neosho plant for repair and was installed in the S-IB-8 stage 5 months later. Engines 101, 102 and 106 also had minor thrust chamber tube separations but were deemed acceptable.

Following inspection and review of firing results the S-IB-6 stage was removed from the test tower on 8 July and shipped from MSFC on the same day, arriving back

at MAF on 13 July. Post-static checkout was completed on 13 October and the stage departed MAF for KSC on 13 December 1966 aboard the barge Palaemon. After arrival at KSC on 18 December 1966 the stage was stacked on pad 37B around 22 January 1967 in preparation for the first LM flight. The S-IVB-206 stage was stacked on top on 23 January. However, as a result of the Apollo 1 fire on 27 January 1967 the Apollo schedule was disrupted and the S-IB-6 stage was de-stacked in early March 1967. It was loaded onto the barge Promise on 31 March, which departed Cape Kennedy on 3 April and returned to MAF, arriving on 10 April 1967, and placed in storage.

The stage was removed from storage on 16 October 1967 for modification and rework. Post-storage checkout was performed between 21 November 1967 and 3 January 1968. The stage was placed in environmental storage at MAF between 18 December 1968 and 18 October 1971. Subsequently, in preparations for its role in the Skylab program the stage underwent modifications and rework, which were completed on 12 February 1972. Post-storage and post-modification checkout started 2 days later and ran through to 6 June 1972.

**Second outboard engine being installed
in the S-IB-6 stage at Michoud** (1.1966)

S-IB-6 stage leaving Michoud (19.5.1966)

The S-IB-6 stage in storage at Michoud (2.1968) 6864131

The CSM-116 arrived at KSC on 19 July 1972 on board the Super Guppy. The S-IB-6 stage departed MAF for KSC once again, on 17 August 1972 on board the barge Orion, arriving on 22 August. The S-IVB-206 stage had arrived at KSC on 24 June 1971 and was removed from storage on 17 April 1972. The S-IU-206 arrived on 22 August. The S-IB-6 stage was transported to the transfer aisle of the VAB, where the eight fins were attached. The S-IB-6 stage was stacked atop the milk stool on ML-1 in High Bay 1 of the VAB on 31 August. The S-IVB-206 stage was stacked on top on 5 September, followed by the IU two days later. A dummy CSM (BP-30) was added together with a stub section of the LES on 8 September, and the stack was rolled out from the VAB to pad 39B on 9 January 1973.

The SA-206 vehicle (with BP CM and stub LES) at pad 39B (9.1.1973) KSC-73PC-10

A propellant loading and All-Systems Test was completed on 30 January 1973. After a series of fit checks the stack was rolled back to the VAB on 2 February. On 9 February 1973 the CSM-116 was mated to SLA-6A and on 20 February the assembly was moved from the O&C building to the VAB. On 20 February BP-30 was de-stacked followed by the installation of CSM-116 on the following day. The BP-30 CM ultimately found a home at the top end of the restored Saturn V in the Saturn V center at KSC after temporarily being located in Fort Worth. The full LES was installed on 24 February before the rocket's final roll back to pad 39B on 26 February 1973. The Flight Readiness Test was conducted on 5 April. RP-1 fuel was loaded on 23 April, and the Countdown Demonstration Test (wet) took place on 3 May 1973. The dry CDDT took place the following day. The rocket was due to be launched on 15 May but the launched was delayed following the problems experienced by the Skylab space station during its launch. Lightning struck the ML-1 lightning mast at 1724 EDT on 24 May but testing showed that there was no damage to the spacecraft.

The S-IB-6 stage formed the first stage of the AS-206 vehicle which launched the first manned crew to the Skylab space station at 0900:00 EDT on 25 May 1973 on board Skylab 2.

S-IB-7

Summary

Stage originally manufactured and tested in 1966. Stored for 5 years before being launched in 1973 as the first stage of the second manned Skylab vehicle.

Engines

The initial engine configuration was:

Position 101: H-7077
Position 102: H-7078
Position 103: H-7079 – removed from the S-IB-7 stage prior to the static firing and installed on the S-IB-8 stage
Position 104: H-7080
Position 105: H-4073
Position 106: H-4074
Position 107: H-4075
Position 108: H-4076

Spare engine: H-7076 – utilized in place of H-7079

The static firing engine configuration was:

Position 101: H-7077 – removed from S-IB-7 and

transferred to RS-27 program
Position 102: H-7078
Position 103: H-7076
Position 104: H-7080 – removed from S-IB-7 and
transferred to RS-27 program
Position 105: H-4073 – removed from S-IB-7, installed
on S-IB-12 and then transferred to RS-27 program
Position 106: H-4074
Position 107: H-4075
Position 108: H-4076

Spare engine: H-7085 – utilized in place of H-7077
Spare engine: H-7074 – utilized in place of H-7080
Spare engine: H-4078 – utilized in place of H-4073

The final flown engine configuration was:

Position 101: H-7085
Position 102: H-7078
Position 103: H-7076
Position 104: H-7074
Position 105: H-4078
Position 106: H-4074
Position 107: H-4075
Position 108: H-4076

Stage manufacturing

Installation of the original engines in the stage was
completed on 11 April 1966.

Stage testing

During the Post Manufacturing Checkout, which was
completed on 12 July 1966, there was a thrust chamber
leak at the aspirator band on engine # 103, H-7079. As
a result that engine was removed from the stage on 13
July and a replacement engine, H-7076, was installed in
the stage on 23 July. The S-IB-7 stage left Michoud on
4 August on board the barge Poseidon, arriving at
MSFC on 11 August. The stage was installed in the Static Test Tower East at MSFC.

The stage was subject to the usual two static firings, the
first of which, SA-38, took place at 1659 CDT on 1
September 1966 and lasted 35 seconds. The full duration firing, SA-39, took place at 1650 CDT on 13 September 1966 and lasted 140 seconds. The official cumulative firing time for the outboard engines was 175.3
seconds, and 172.6 seconds for the inboard engines.

The S-IB-7 stage was shipped from MSFC back to
Michoud on 20 September, on board the barge Palaemon, arriving 5 days later. Post-static modifications and
repairs were started on 26 September. During the
course of Post Static Checkout at Michoud, which had
begun on 1 November, a number of engine problems

were discovered. Engine 104, H-7080 was removed on
14 November 1966 and replaced by engine H-7074,
which was installed in the stage 2 days later after having arrived in Michoud on 12 September 1966. The
removed engine H-7080 was discovered to have a number of pieces of Teflon behind the injector plate. In
addition analysis of test data revealed that the engine
performance had changed from engine-level testing to
stage-level testing as a result of the contamination.

With the Post Static Checkout being completed on 27
December 1966 it was decided to remove engine 105,
H-4073, the following day as it was believed that this
engine could also have suffered from the ingress of
Teflon shards that had been determined to have origi-
nated from one of the LOX tanks. Inspection also
revealed that this engine had 5 turbine blades manufactured from the wrong material, similar to those discovered on the S-IB-8 stage. Six months elapsed before the
replacement engine, H-4078 was installed in the S-IB-
7 stage. That engine itself had been removed from
another stage, S-IB-8 on 21 November 1966 after it had
defective turbine blades discovered. After recertifica-
tion firings on 2 February 1967 engine H-4078 arrived
at Michoud on 23 March 1967 and was installed in the
S-IB-7 stage on 10 June 1967, 2 days after completing
Post Modification Checkout.

Engine # 102, H-7078, was removed from the stage at
the conclusion of Post Static Checkout and returned to
Rocketdyne for a turbine wheel replacement after the
discovery of 3 blades of incorrect material. Rocketdyne
performed the repair, undertook repeat test firings on
the engine on 13 January 1967 and shipped the engine
back to Michoud on 28 January 1967, where it arrived
2 days later. The engine was re-installed in its original
102 position on 16 February 1967, 2 days after completing Post Modification Checkout.

Because of delays to the program the S-IB-7 stage was
placed in long term storage at Michoud on 18 July
1967. Periodic health checks were performed, and during one of these a problem arose with engine # 101, H-
7077. Socket head capscrew contamination was discovered. Consequently this engine was removed from the
S-IB-7 stage and replaced with engine H-7085, which
itself had been removed from the S-IB-8 stage on 15
December 1966. After revalidation test firings on 5 January 1967 H-7085 was delivered to Michoud on 14
August 1967 and installed in the S-IB-7 stage on 16
October 1968. On 20 April 1967 CCSD moved the S-
IB-7 stage into the functional checkout complex for a
re-verification checkout.

The S-IB-7 stage entered official long-term environmental storage at Michoud on 14 January 1969. Unlike
the previous stage, S-IB-6, this stage was not delivered
to KSC at the time the Apollo program was put on hold

because of the Apollo 1 fire. After 3 years the stage was removed from environmental storage on 31 January 1972. Post Storage and Post Modification Checkout started on 24 May 1972 and continued until 28 August 1972.

The S-IVB-207 stage arrived at KSC on 26 August 1971 and the CSM-117 arrived on 1 December 1972. The S-IB-7 stage was shipped from Michoud on 24 March 1973, arriving at KSC on 30 March 1973. The S-IB-7 was erected on ML-1 on 28 May 1973, followed by the S-IVB-207 later on the same day. The Instrument Unit for this mission S-IU-208 (which was out of sequence) arrived at KSC on 8 May 1973 and was erected on the launch vehicle on 29 May. Erection of the SLA-23 and CSM-117 followed.

The launch vehicle was transferred to the launch pad 39B on 11 June and the Flight Readiness Test was conducted on 29 June. RP-1 fuel was loaded into the S-IB-7 stage on 11 July and the wet CDDT was performed on 20 July.

The S-IB-7 formed the first stage of the AS-207 vehicle that launched the second manned crew to the Skylab space station at 0710:50 EDT on 28 July 1973.

S-IB-7, S-IB-9, S-IB-5 and S-IB-6
in final assembly at Michoud (10.1967) 6760550

S-IB-8

Summary

Stage originally manufactured and tested in 1966. Severe damage to the H-1 engine turbine blades during static test due to the blades being made of the wrong material. Stage stored for 5 years before being launched in 1973 as the first stage of the third manned Skylab vehicle. Included one engine that had been part of test firings on two different stages. Cracked fins replaced at the launch pad.

Engines

The initial build engine configuration was:

Position 101: H-7082
Position 102: H-7083
Position 103: H-7084 – removed prior to the static firings and replaced by H-7081
Position 104: H-7085
Position 105: H-4077
Position 106: H-4078
Position 107: H-4079
Position 108: H-4080

Spare engine – H 7081 – installed in place of H-7084

The first stage static firing engine configuration was:

Position 101: H-7082
Position 102: H-7083
Position 103: H-7081
Position 104: H-7085
Position 105: H-4077
Position 106: H-4078 – removed from the S-IB-8 stage after the first static firing and later used on S-IB-7
Position 107: H-4079
Position 108: H-4080

Spare engine: H-4071 – utilized in place of H-4078

The second stage static firing engine configuration was:

Position 101: H-7082
Position 102: H-7083 – removed from S-IB-8 and transferred to the RS-27 program
Position 103: H-7081
Position 104: H-7085 – removed from S-IB-8 and used on S-IB-7
Position 105: H-4077
Position 106: H-4071
Position 107: H-4079
Position 108: H-4080

Spare engine: H-7079 – utilized in place of H-7083
Spare engine: H-7096 – utilized in place of H-7085

The final flown engine configuration was:

Position 101: H-7082
Position 102: H-7079
Position 103: H-7081
Position 104: H-7096
Position 105: H-4077
Position 106: H-4071
Position 107: H-4079
Position 108: H-4080

Stage manufacturing

The original engines were installed in the stage by 22 June 1966.

Stage testing

During test activities engine # 103 (H-7084) had a problem and was replaced by a spare engine, H-7081. The replaced engine became an R&D spare and was eventually scrapped. The stage completed Post Manufacturing Checkout on 26 September 1966 and S-IB-8 was shipped from Michoud on 17 October, arriving at MSFC on 25 October. The stage was installed in the Static Test Tower East on 25 October and simulated flight tests with flight pressure were performed on 7, 15 and 28 November. The propellant loading test was performed on 8 November. Prior to the first static firing a Rocketdyne welder repaired an external dent on tube 27 of engine # 106 (H-4078). Difficulties were also encountered with this engine # 106 (H-4078) during the initial turbo-pump torque checks. The drive shaft would not rotate and eventually required manual agitation to shake it loose. After this an intermittent clicking sound could still be heard as the turbine was rotated.

The stage underwent the first static firing, designated SA-40, at 1640:04.000 CST on 16 November 1966 with a duration of 35.444 seconds (inboard engines), 35.560 seconds (outboard engines). Cut-off was initiated by the firing panel operator. There were problems with engine # 106 (H-4078), which exhibited an out-of-specification thrust of 184,000 pounds, which resulted in that engine being removed from the stage on 21 November. Inspection revealed severe damage to the first stage turbine blades. A count revealed 42 blades missing from the first stage wheel. A reduction analysis of the data revealed that damage to the engine occurred during the first ten seconds of the test.

The replacement engine for H-4078 was H-4071, which had seen two stage firings when installed on S-IB-6, but had been removed from that stage on 29 June 1966. Following engine-level revalidation firings at Rocketdyne's Neosho plant on 14 July 1966 H-4071 arrived at Michoud on 12 August 1966. The engine was dispatched to MSFC on 17 November, a day after the problem was encountered during the first stage firing of S-IB-8. The engine completed Post Modification Checkout en-route to MSFC, where it arrived on 20 November. It was installed in the # 106 location on 23 November; 2 days after H-4078 had been removed. None of the other engines needed re-orificing after the short firing.

This cleared the way for the second, full duration firing of the stage (SA-41) which took place at 1640:54.000

CST on 29 November 1966 for a duration of 142.656 seconds (inboard engines), 145.352 seconds (outboard engines). The cumulative total firing times for the outboard engines was 181.1 seconds, and 178.1 seconds for the inboard engines. The inboard engine cut-off signal was initiated by the switch selector. This was 3.2 seconds after the LOX low level sensor 3 in tank O-4 was uncovered. Outboard engine cut-off was initiated by dropout of two engine # 103 (H-7081) Thrust OK pressure switches as a result of LOX depletion. Turbine inlet pressure on engine # 107 (H-4079) was below normal throughout test SA-41. Removal of the solenoid calibration valve revealed fuel and carbon residue clogging the inlet screen.

The stage was removed from the test stand on 8 December 1966 and shipped from MSFC on the same day, arriving back at Michoud on 14 December 1966. After the stage was received at Michoud it entered Post Static Checkout.

Analysis by Rocketdyne of the first stage turbine wheel from the engine that failed in the short firing (H-4078) revealed the presence of type 316 stainless steel turbine blades, instead of the specified Haynes Stellite 21 material. X-ray analysis by Rocketdyne indicated that stainless steel blades were also present in the first stage turbine wheels of engines # 102 (H-7083), 11 blades, 103 (H-7081), 6 blades, 104 (H-7085), one blade and 105 (H-4077), 33 blades. These four engines were removed from the stage, returned to Neosho, where replacement first stage turbine wheels were installed and the H-1 engines static fired again by Rocketdyne. Upon return to Michoud it was intended that they would all be installed in their original positions on the S-IB-8 stage.

Engine 104 (H-7085) was removed on 15 December. Engines 102 (H-7083), 103 (H-7081) and 105 (H-4077) were also removed in mid December. The latter two engines each had a turbine wheel replaced, were re-verified with single engine tests and were re-installed in

Test firing of an S-IB stage at MSFC (Mid 1960s)

S-IB-8 in storage at Michoud (4.1968) 6866324

their original locations on the stage. H-7081 was tested on 9 January 1967 and arrived back at Michoud on 23 January before being re-installed in S-IB-8 on 2 February 1967. H-4077 was tested on 29 December 1966 and arrived back at Michoud on 17 January 1967 before being re-installed in S-IB-8 on 26 January 1967.

After engine H-7085 was removed from position # 104, the turbine wheel replaced and the engine re-fired at Neosho, it was decided to transfer that engine to the S-IB-7 stage. This was because the LOX seal cavity had become contaminated and required cleaning. A replacement engine, H-7096, was identified for use in the S-IB-8 stage. This replacement engine was removed from S-IB-11, position 102, on 11 May 1967, prior to that stage's static firing. The engine was acceptable at the time of its removal and was simply being used as a spare. Consequently it did not undergo any further engine-level testing, having been subjected to its final regular acceptance test on 2 August 1966. The engine arrived in Michoud on 19 June 1967 and was installed as the replacement engine in the S-IB-8 stage on 10 July 1967.

The S-IB-8 stage Post Static Checkout was completed on 17 April 1967. The final engine that had been returned to Neosho in December 1966 for a turbine wheel replacement, engine 102 (H-7083), was finally removed from the stage on 20 July 1967 because the LOX seal cavity had become contaminated and required cleaning. The replacement engine, H-7079, had been installed previously on stage S-IB-7, and had been removed from that stage on 13 July 1966 prior to the static firing of the S-IB-7 stage. Engine H-7079 underwent re-verification testing on 24 October 1966 prior to receipt at Michoud on 21 November 1966. This replacement engine was installed in the S-IB-8 stage on 1 August 1967 thus completing the engine set. No further engine exchanges occurred in the following 6 years leading to launch.

The problem of incorrect turbine wheel material was discovered to have affected engines in the Atlas and Thor boosters as well as the Saturn IB. A total of 816 wheel x-rays were examined and 28 of these had blades of mixed materials. Ten of the 28 wheels were in H-1 engines; ten were in Thor and Atlas sustainer engines; 6 were in R&D engines and 2 were scrapped in production.

Only 3 engines (positions 101, 107 and 108) on this stage had remained in place without removal from the original stage installations in June 1966 up to launch in November 1973.

With delays to the Saturn program the S-IB-8 stage entered environmental storage in Michoud on 29 August 1968. It remained there until 22 March 1972. Post Storage and Post Modification Checkout was completed on 16 January 1973. The S-IB-8 stage was shipped from Michoud on 15 June 1973, arriving at KSC on 20 June 1973.

The S-IB-8 stage was erected on the mobile launcher, ML-1, on 31 July 1973. The S-IVB-208 stage, which had arrived at KSC on 5 November 1971, was erected on top of the booster stage on the same day. The out-of-sequence IU, S-IU-207, arrived at KSC on 12 June 1973 and was stacked on the rocket on 1 August 1973. Following this the SLA-24 and CSM-118 were mounted on top of the rocket.

The rocket was transferred to launch pad 39B on 14 August 1973. On 28 August a crack was discovered in Chanel, Upper Outrigger Assembly, Fin Position 4. This was repaired on 4 September 1973. Flight Readiness Tests were conducted on 5 September and 11 October. RP-1 fuel was initially loaded on board on 23 October. Inversion of fuel tanks F3 and F4 occurred during this loading because a vent cover had not been removed. The tanks were reformed two days later. The wet CDDT was performed on 2 November.

The launch was initially scheduled for 10 November but was delayed as all 8 S-IB-8 tail fins had to be replaced. This was after inspection of the tail fins on the rescue booster, S-IB-9, in the VAB revealed three cracks. An inspection of the S-IB-8 stage on 6 November revealed similar cracks in the fin attachment fittings caused by stress corrosion cracking. Cracks were found in all 8 fin attachment fittings with two cracks on seven fins and one crack on the eighth. The longest crack was 1.5 inches. On 7 November the RP-1 was drained in preparation for replacement of the tail fins. Work on removing the first fin began at 1433 EST on 8 November and was completed 35 hours later. More cracks were discovered in seven of the eight S-IB/S-IVB interstage reaction beams. It was decided that these new cracks would be acceptable with no rework. However the task of replacing the fins continued until completed at 0704 EST on 13 November. The RP-1 was re-loaded on the following day.

The S-IB-8 stage finally was launched as the first stage of the third and final manned Skylab flight, AS-208 Skylab 4, at 0901:23 EST on 16 November 1973.

S-IB-9

Summary

Stage subject to four acceptance firings. Stage acted as stand-by rescue vehicle for the final manned Skylab mission and for the Apollo Soyuz flight two years later. Rolled out to the launch pad for the Skylab potential rescue. Currently on display at KSC.

Engines

The initial engine configuration was:

Position 101: H-7086
Position 102: H-7087
Position 103: H-7088
Position 104: H-7089
Position 105: H-4082
Position 106: H-4083 – removed from the stage and replaced by H-4081
Position 107: H-4084
Position 108: H-4085

Spare engine: H-4081 – utilized in place of H-4083

The subsequent engine configuration was:

Position 101: H-7086 – removed from the S-IB-9 stage and installed on the S-IB-11 stage
Position 102: H-7087
Position 103: H-7088
Position 104: H-7089
Position 105: H-4082
Position 106: H-4081 – removed from the S-IB-9 stage and later used as a Flight Worthiness Verification engine in support of the S-IB-9 stage
Position 107: H-4084
Position 108: H-4085

Spare engine: H-7090 – utilized in place of H-7086
Spare engine: H-4086 – utilized in place of H-4081

The stage static firing engine configuration was:

Position 101: H-7090 – removed from the S-IB-9 stage after the static firing and transferred to the RS-27 program
Position 102: H-7087

Cracked fins on the S-IB-9 stage being replaced in the VAB at KSC (Mid 9.1973)

Position 103: H-7088
Position 104: H-7089
Position 105: H-4082
Position 106: H-4086
Position 107: H-4084
Position 108: H-4085

Spare engine: H-7110 – utilized in place of H-7090

The final engine configuration was:

Position 101: H-7110
Position 102: H-7087
Position 103: H-7088
Position 104: H-7089
Position 105: H-4082
Position 106: H-4086
Position 107: H-4084
Position 108: H-4085

Spare engine: H-7113 (not used, transferred to KSC)
Spare engine: H-4109 (not used, transferred to KSC)

Currently the stage on display at the KSC Rocket Garden has no real engines attached.

Stage manufacturing

The engines were installed in the S-IB-9 stage between 19 August and 12 September 1966.

Stage testing

During initial testing a problem was found with engine # 106 (H-4083) and the engine was removed and replaced by spare engine H-4081. The replaced engine was assigned to the 205,000 pounds thrust re-qualification program and was eventually scrapped at the completion of testing.

Post Manufacturing Checkout was completed in Michoud on 20 December 1966. During the checkout

The SA-209 Skylab rescue vehicle being rolled back into the VAB at KSC (2.1974) KSC-73P-682

The SA-209 vehicle on display at KSC (2006)

special inspections were made of the turbine wheel material following the detection of incorrect material in some blades on S-IB-8 engines. Consequently two engines were identified with incorrect blades and the engines were replaced on the stage. Engines 101 (H-7086), 5 incorrect blades, and 106 (H-4081), 6 incorrect blades, were removed from the stage on 20 December. Engine H-7086 was replaced by H-7090 which had completed engine-level testing on 31 May 1966 and had arrived at Michoud on 17 June 1966. This replacement engine was installed in the stage on 5 January 1967. Engine H-4081 was replaced by H-4086 which had completed engine-level testing on 20 May 1966 and arrived at Michoud on 17 June. It was installed in the S-IB-9 stage on 30 December 1966. The S-IB-9 stage departed Michoud on board the barge Palaemon on 19 January 1967, arriving at MSFC on 25 January.

The stage was installed in the Static Test Tower East on 27 January 1967. Simulated flight tests were performed on 2 and 23 February and 2 and 6 March. During the course of routine preparations for static test, three thrust OK pressure switches and two associated electrical cables, four auxiliary hydraulic pump motors and one ignition monitor valve were replaced because of malfunction or evidence of damage. The propellant loading test was performed on 15 February 1967.

Prior to the first static firing Union Carbide personnel performed an inspection of the turbine wheel blades and determined that the blades were made from the specified Haynes Stellite material. Also at this time excessive leakage past the gas generator control valve

LOX poppet seat was detected on engine # 101 (H-7090). The gas generator was disassembled by Rocketdyne personnel at the MSFC component repair area. A white powdery substance was found imbedded in the seat. The seat was cleaned, reinstalled and checked.

The stage underwent 4 static firings in the Static Test Tower East at MSFC. The first attempt at the short trim run took place on 24 February 1967 at 1708:05.985 CST. However this firing (SA-42) was aborted after 13.252 seconds (inboard engines), 13.528 seconds (outboard engines), of the planned 35 seconds because of a power failure in the Beckman digital data acquisition system. This condition caused the indication of a low lube oil pressure, below the redline limit, and a cut-off command was automatically given.

Three days later on 27 February at 1641:44.000 CST the S-IB-9 stage was successfully fired (SA-43) for the required 35.324 seconds (inboard engines), 35.440 (outboard engines). Cut off was initiated by the firing panel operator. All engines produced thrust within the specified range of 205,000 +/- 3,000 pounds and no re-orificing was required. The Main Fuel Valve on engine

The SA-209 vehicle on display at KSC (2006)

104 (H-7089) was replaced prior to test SA-44 due to excessive leakage.

The first attempt at the full duration firing (SA-44) was also unsuccessful. This firing at 1545:43.000 CST on 7 March was terminated at the time of ignition giving an official duration of 3.080 seconds (inboard engines), 3.356 seconds (outboard engines). Cut-off was due to a loose connection in the Beckman Digital Data System.

However, the stage was turned around in a matter of hours and the full duration firing (SA-45) was performed successfully for 142.400 seconds (inboard engines), 145.448 seconds (outboard engines) at 1821:57.000 CST on the same day. Inboard engine cut-off signal was initiated by the switch selector, and the outboard engine cut-off signal was initiated by dropout of engine # 103 Thrust OK pressure switch as a result of LOX depletion. The official cumulative totals were 197.8 seconds for the outboard engines and 194.0 seconds for the inboard engines. All engines performed satisfactorily during the long firing and produced thrust values within the specified limits, except for engine # 103 (H-7088). This engine exhibited a high thrust value of 208,430 pounds, but a decision was made not to re-orifice the engine. This engine also accumulated 43 counts of rough combustion during ignition of test SA-45. Post test SA-45 inspection showed that the outboard engine aspirator lips were burned on the inboard quadrant of engines 103 (H-7088) and 104 (H-7089), and a 0.025 inch deep dent was noted on the inside of engine 104. Repairs to the dent were made by Rocketdyne.

The stage was removed from the Static Test Tower East on 14 March and loaded onboard the barge later that day. The S-IB-9 stage was shipped from MSFC in the barge Palaemon on 15 March 1967, arriving at Michoud on 20 March. Post Static Checkout was completed on 26 July 1967 and the stage entered a period of storage at MAF on 3 August 1967. It formally entered Environmental Storage on 12 February 1969 and stayed there for over three years until it was removed on 6 July 1972 for modifications and rework, which were completed on 22 November 1972. During this period engine 101 (H-7090) was removed and replaced. The new engine, H-7110, had not been fired since 11 Jul 1967 and had remained in storage at Michoud since it arrived on 28 August 1967. It underwent a new post-modification checkout on 21 August 1972 and was installed in the S-IB-9 stage on 27 October 1972.

Post-storage and post-modification checkout of the stage was completed on 28 March 1973. The S-IB-9 stage left Michoud on 14 August 1973 and arrived at KSC on 20 August. During the third and final manned Skylab mission the S-IB-9 stage acted as the first stage of a potential rescue mission, if needed. As such, it was mated to the S-IVB-209 stage, SLA-22 and CSM-119

in the VAB. However, prior to that, in early November 1973, cracks were detected in some of the 8 stabilizing fins on the S-IB-9 stage. This lead to an inspection of the S-IB-8 vehicle, which was ready for launch on pad 39B. Cracks were also detected in some of the fins on this launch vehicle and the lift-off was delayed some days whilst the fins were replaced in situ on the pad. The cracked fins on the S-IB-9 stage were also replaced in the VAB prior to the stage stacking.

Later, following the successful launch of the final Skylab vehicle the SA-209 stack, mounted on the milk-stool, was rolled out to pad 39B on 3 December 1973. A Flight Readiness Test was performed on 17 December 1973. It remained there until the completion of the final Skylab mission in February 1974, whereupon the stack was rolled back to the VAB and disassembled, as it was no longer needed for a potential rescue mission. The S-IB-9 stage was placed in environmental storage at KSC on 17 April 1974. It was joined in storage by spare engine, H-7113 on 19 November 1974. Following removal of the stage fin assemblies from the environmental enclosure the S-IB-9 stage was returned to environmental storage.

In 1975 it was again readied for potential launch, this time as the back-up Apollo Soyuz launch vehicle. However, unlike the Skylab rescue mission, the S-IB-9 stage was not assembled into a complete Saturn IB stack or transferred to the launch pad. After the success of the Apollo Soyuz mission in 1975 the S-IB-9 stage was no longer needed as a back up for that mission and has ever since been on display in a horizontal configuration in the KSC Rocket Garden, together with the S-IVB-209 upper stage and payload. The 8 H-1 engines were removed in 1994 and replaced by red engine covers.

S-IB-9 stage is loaded on board the Palaemon at Michoud
(19.1.1967)

S-IB-10

Summary

This stage was originally built and tested in 1966-67 but did not fly until 1975 when it provided the first stage on the final Apollo-Saturn mission, the joint US-USSR Apollo Soyuz Test Project. The full duration stage acceptance test firing was witnessed by Vice President Hubert Humphrey in May 1967.

Engines

The initial engine configuration was:

Position 101: H-7091
Position 102: H-7092 – removed from the S-IB-10 stage and installed on the S-IB-11 stage
Position 103: H-7093
Position 104: H-7094
Position 105: H-4087
Position 106: H-4088
Position 107: H-4089
Position 108: H-4090

Spare engine: H-7099 – utilized in place of H-7092

The stage static firing engine configuration was:

Position 101: H-7091
Position 102: H-7099
Position 103: H-7093
Position 104: H-7094
Position 105: H-4087
Position 106: H-4088 – removed from the S-IB-10 stage after the stage static firing and transferred to the RS-27 program
Position 107: H-4089
Position 108: H-4090

Spare engine: H-4104 – utilized in place of H-4088

The final flown engine configuration was:

Position 101: H-7091
Position 102: H-7099
Position 103: H-7093
Position 104: H-7094
Position 105: H-4087
Position 106: H-4104
Position 107: H-4089
Position 108: H-4090

Spare engine: H-7112 (not used)
Spare engine: H-7113 (not used, transferred to S-IB-9 spare, transferred to KSC)
Spare engine: H-7097 (not used, transferred to FWV

Vice President Humphrey and Dr von Braun at MSFC to witness the S-IB-10 SA-47 firing (22.5.1967) 6755828

engine, transferred to EFL)
Spare engine: H-4108 (not used)
Spare engine: H-4109 (not used, transferred to S-IB-9 spare, transferred to KSC)

Stage manufacturing

Stage manufacturing was started on 6 September 1966. The 8 H-1 engines were originally installed in the S-IB-10 stage between 25 October and 5 December 1966 at Michoud.

Stage testing

During Post-Manufacturing Checkout at Michoud a special inspection of the material of the H-1 engine turbine wheel blades was made following the discovery of blades manufactured from incorrect material on the S-IB-8 stage. Engine # 102 (H-7092) was discovered to have 2 incorrect turbine wheel blades. Consequently that engine was removed from the stage on 13 December 1966 and the replacement engine, H-7099, which had arrived at Michoud on 10 October 1966, was installed in the S-IB-10 stage on 20 December 1966. Post-Manufacturing Checkout was completed on 7 March 1967 and the stage left Michoud on board the barge Palaemon on 31 March 1967, arriving at MSFC on 7 April.

The stage was installed in the MSFC Static Test Tower East on 10 April 1967. Simulated flight tests were performed on 27 April, 8 and 21 May. The propellant system functioned satisfactorily during the propellant loading test on 28 April 1967. The S-IB-10 stage was subjected to the standard 2 acceptance firings in the Static Test Tower East at MSFC. The first firing (SA-46) took place at 1705:31.000 CDT on 9 May 1967 and lasted 35.308 seconds (inboard engines), 35.424 seconds (outboard engines). Cut-off was initiated by the firing panel operator.

The full duration firing (SA-47) was originally scheduled for 17 May but was delayed 5 days to permit the Vice President to view the firing. It finally took place at 1733:55.000 CDT on 22 May 1967 and lasted 142.712 seconds (inboard engines), 145.712 seconds (outboard engines). As planned the firing was witnessed by Vice President Hubert H Humphrey, who was touring MSFC that day. The official cumulative firing times were 181.1 seconds for the outboard engines and 178.0 seconds for the inboard engines. Inboard engine cut-off was initiated by the switch selector. Outboard engine cut-off was initiated by the backup timer.

All stage systems produced satisfactory results. All engines produced thrust values within the specified 205,000 +/- 3,000 pounds except engine # 108 (H-4090) during test SA-46. Data indicated that the thrust value of this engine was 209,645 pounds. The orifice was replaced and a nominal performance was recorded in the firing SA-47.

A special LOX seal cavity swab test was performed on S-IB-10 because leak-tek solution had been found in the LOX seal cavity on several S-IB-11 engines at Michoud. However, no contamination was found in this stage. A turbine wheel inspection was performed by Union Carbide personnel prior to the short firing. The blades were found to be made from the specified Haynes Stellite material. Two special heat shield panels installed on the stage showed no damage following the tests.

The S-IB-10 stage was removed from the static test tower on 7 June. It left MSFC on board the barge Palaemon on 8 June 1967 and arrived back at Michoud on 13 June. The stage was put into storage at Michoud, on 30 August 1967, without Post-Static Checkout being performed. Formal environmental storage was started on 10 April 1969 where the vehicle remained until 30 October 1972. The stage then underwent modification and rework which was completed on 13 April 1973. During this time a problem with engine 106 (H-4088) was detected. The engine was removed from the stage and replaced with engine H-4104, which had not been fired since 8 May 1967. That engine had been in storage at Michoud since 3 July 1967 and was installed in

the S-IB-10 stage on 16 February 1973. Engine H-7097 was assigned as a Flight Worthiness Verification engine for this stage on 12 June 1973. Post-storage and post-modification checkout was completed on 23 July 1973. Between 26 and 29 November 1973 the flight heat shields were removed.

The stage, together with the FWV engine, were shipped from Michoud on 19 April 1974 and arrived at KSC on 24 April 1974. The S-IB-10 stage was placed in environmental storage at KSC on 29 April where it remained until 3 December 1974. On 9 May 1974 it was joined in storage by flight spare engines, H-7113 and H-7097.

On 13 January 1975 the S-IB-10 stage was erected on the mobile launcher. The second stage was stacked the following day, the IU and boilerplate spacecraft on 16 January. On 19 February hairline cracks in 2 of the 16 hold down fittings were discovered. By the end of February cracks in 6 other fins were detected. Consequently it was decided to replace all the fins as they had been on the last Skylab launch vehicle and the rescue vehicle. The CSM-111 was mated to the SLA-18 on 5 March. Between 11 and 19 March all the fins on the S-IB-10 stage were replaced. After removal of the boilerplate spacecraft the flight spacecraft was erected on the launch vehicle on 19 March 1975.

On 24 March 1975 the AS-210 vehicle was rolled to launch pad 39B. On top of the booster stage was the S-IVB-210 stage, the S-IU-210, SLA-18, DM-2 and the CSM-111. The CDDT took place on 3 July 1975. The vehicle was launched at 1550:00 EDT on 15 July 1975 with the S-IB-10 stage forming the first stage of the AS-210 launch vehicle that orbited the final manned Apollo capsule used in the Apollo Soyuz Test Project.

S-IB-10 stage leaving Michoud (31.3.1967)

S-IB-11

Summary

This stage underwent an unprecedented seven static firings. These included special tests of combustion instability. Engine explosion in one test firing due to LOX seal leak. Stage never flown and currently on display at the Alabama Welcome Center on the I-65.

Engines

The initial engine configuration was:

Position 101: H-7095 – removed from the S-IB-11 stage position 101 and transferred to the S-IB-11 stage position 102
Position 102: H-7096 – removed from the S-IB-11 stage and installed on the S-IB-8 stage
Position 103: H-7097 – removed from the S-IB-11 stage and became a Flight Worthiness Verification Vehicle for the S-IB-12 and S-IB-10 stages before ending up at Edwards Field Laboratory on 13 December 1974
Position 104: H-7098 – removed from the S-IB-11 stage, installed in the S-IB-12 stage and finally transferred to the RS-27 program
Position 105: H-4092
Position 106: H-4093
Position 107: H-4094
Position 108: H-4095

Spare engine: H-7092 – utilized in place of H-7095
Refurbished engine: H-7095 – utilized in place of H-7096
Spare engine: H-7086 – utilized in place of H-7097
Spare engine: H-7102 – utilized in place of H-7098

The stage static firing engine configuration for the first stage firing was:

S-IB-11 on board the Palaemon en route to MSFC from Michoud (20.10.1967) 6760498

Position 101: H-7092 – temporarily removed until the 6th firing
Position 102: H-7095
Position 103: H-7086
Position 104: H-7102
Position 105: H-4092
Position 106: H-4093
Position 107: H-4094 – temporarily removed until the 6th firing
Position 108: H-4095

R&D engine: H-T6-B temporarily installed in place of H-7092
R&D engine: H-4067 temporarily installed in place of H-4094

The stage static firing engine configuration for the second to fifth stage firings was:

Position 101: H-T6-B – R&D engine used for bomb test

Re-installation of the H-7095 engine in the S-IB-11 stage at Michoud (14.7.1967) 6757861

The S-IB-11 stage being off-loaded from the Palaemon at the MSFC dock (27.10.1967) 6760211

**The S-IB-11 stage being off-loaded
from the Palaemon at the MSFC dock** (27.10.1967) 6760213

The S-IB-11 stage arrives at the MSFC test stand
(27.10.1967) 6760216

Position 102: H-7095
Position 103: H-7086
Position 104: H-7102
Position 105: H-4092
Position 106: H-4093
Position 107: H-4067 – R&D engine used for bomb test
Position 108: H-4095 – removed from the S-IB-11 stage following fifth stage firing, turbo pump and propellant ducts removed, used for rebuilding engine HT6B at MSFC

Spare engine: H-4091 – utilized in place of H-4095

The stage static firing engine configuration for the sixth and seventh stage firings was:

Position 101: H-7092
Position 102: H-7095
Position 103: H-7086
Position 104: H-7102
Position 105: H-4092
Position 106: H-4093

Position 107: H-4094
Position 108: H-4091

Spare engine: H-7112 (not used)
Spare engine: H-4108 (FWV with this stage)
Spare engine: H-4107 (FWV with this stage after FWV with S-IB-13)

This was the last reported engine configuration in the S-IB-11 stage and it is assumed that these are the engines currently installed in the stage.

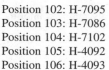

The S-IB-11 stage is transported to the MSFC test stand
(27.10.1967) 6760210

The S-IB-11 stage is installed in the test stand at MSFC
(1.11.1967) 6760212

Removal of test engine from S-IB-11 stage at MSFC
(3.1968) 6864273

Stage manufacturing

The H-1 engines were originally installed in the S-IB-11 stage in January and February 1967, with the last engine in place on 20 February.

During the course of engine checkout on the stage in May 1967 LOX seal cavity contamination problems were discovered in all four outboard engines resulting in their removal from the stage. Engine 101 (H-7095) was the first to come out on 8 May. It was replaced by H-7092, which had previously been installed in the S-IB-10 stage and was removed from that stage on 13 December 1966. It underwent re-verification firings at

The S-IB-11 plus upper stage at the I-65 Alabama Welcome Center (2004)

Rocketdyne on 22 February 1967 and arrived back at Michoud on 23 March. After Post Modification Checkout was completed on 5 June the engine was installed in the S-IB-11 stage on 8 June.

The 102 engine (H-7096) was removed on 11 May 1967 and replaced with engine H-7095, which itself had been located on the same stage in the 101 position before removal on 8 May 1967. Before re-installation in the S-IB-11 stage engine H-7095 did not need to undergo any additional firings; its last engine-level test having been on 28 July 1966. After checkout the engine

Shuttle Enterprise heads south along MSFC's Rideout Road past the horizontal SA-211 stages (15.3.1978) 7887857

H-1 engines on the S-IB-11 stage at the I-65 Alabama Welcome Center (2004)

**The S-IB-11 plus upper stage at the I-65
Alabama Welcome Center** (2004)

tamination. It was completed on 27 September 1967 and the stage was shipped from Michoud on 20 October on board the barge Palaemon, arriving at MSFC on 27 October.

Stage testing

The S-IB-11 stage was installed in the Static Test Tower East at MSFC on 1 November 1967. A simulated flight test was performed on 16 November followed by a propellant loading test on the following day. During this pre-firing checkout Union Carbide personnel inspected the turbine wheel blades on all engines for proper material. However the blades were found to be made from the specified Haynes Stellite material. Further simulated flight tests on 28 November and 18 December cleared the way for the static firings.

The stage was subjected to an unprecedented seven static firings at MSFC which stretched over a four month period. A series of five combustion instability tests were scheduled to investigate the combustion instability self-damping characteristics of the H-1 engine in a stage configuration. The special series of tests was required to gain confidence in the ability of the stage system to regain or return to stable operation after the combustion process in two of the eight engines had been momentarily driven unstable by explosive charges ("bombs") in the operating combustion chamber of these engines.

This series of tests was found necessary when it was determined for the first time in mid-1967 for single engine, production, injector stability testing at the Rocketdyne Neosho Static Test Facility that the 205,000 pounds thrust H-1 engine's ability to regain stable operation after being synthetically disturbed could be adversely affected by a combination of engine, propellant feed system and structural dynamic characteristics. This new finding made it imperative to conduct system stability tests on flight system hardware to regain confidence that the real system would not react unfavorably to unstable engine operation.

The first, short duration firing (SA-48) took place at 1640:21.000 CST on 19 December 1967 with a duration of 35.508 seconds (outboard engines) and 35.392 seconds (inboard engines) as planned. Cut-off was initiated by the firing panel operator. During this test the turbine inlet pressure measurements at engines 102 and 106 were lost due to erroneous calibration. Engine 102 exhibited a high thrust (210,141 pounds) and was re-orificed following this test. Engine 104 (208,517 pounds) and engine 106 (208,616 pounds) also exhibited high thrust during this test, but it was decided to continue without further modifications to these engines.

In preparation for the combustion instability testing engines 101 (H-7092) and 107 (H-4094) were replaced

arrived back at Michoud on 5 June 1967 and was installed in the 102 position on 14 July, just over 9 weeks after it had been removed from the same stage.

The third engine to be removed was engine 103 (H-7097) which came out on 12 May, to be replaced by H-7086, which had been removed from the S-IB-9 stage on 20 December 1966. H-7086 completed engine-level re-verification firings on 25 January 1967 and arrived back at Michoud on 13 February. After completing Post Modification Checkout on 9 June 1967 it was installed in the S-IB-11 stage on 13 June 1967.

The final engine to be removed during May 1967 was engine 104 (H-7098) which came out on 12 May, the same day as engine H-7097 was removed from the 103 position. The replacement engine was H-7102 that had not been associated with any previous stage. This engine had last been test fired by Rocketdyne on 27 October 1966, and had arrived at Michoud on 21 November 1966. The engine was installed in the S-IB-11 stage on 8 June 1967, one day after completing Post Modification Checkout.

Post Manufacturing Checkout of the S-IB-11 stage was resumed on 15 August, following a suspension in July to allow inspection of the H-1 engines for possible conby research and development engines H-T6-B and H-4067 respectively. It had been demonstrated that these test engines had an inherent ability to sustain stable operation as demonstrated in a series of stability tests with single engine firings on the Power Plant Test Stand at MSFC. A further stage modification was that the thrust structure above engine # 107 was strengthened. With this configuration a simulated flight test was performed on 24 January 1968. A Rocketdyne 99-651932 bomb for the instability test was installed in the R&D engine at position 107 only at this time.

The first firing with this development engine configuration (SA-49) took place at 1707:45.013 CST on 25 January 1968, with a duration of 15.644 seconds (outboard engines) and 15.528 (inboard engines), with cut-off initiated by the firing panel operator. The bomb in engine 107 was detonated 2.402 seconds after ignition command, with a period of instability lasting 26ms. During this test engine 108 (H-4095) exhibited high thrust (208,565 pounds) but the disposition was to use-as-is.

In preparation for the second instability test firing bombs were placed in both R&D engines. In addition the thrust structure above engine 101 was strengthened in a similar manner to that done at position 107 prior to test SA-49. A simulated flight test was performed on 5 February. The firing, (SA-50), started at 1640:02.000 CST on 6 February 1968, lasted 15.576 seconds (outboard engines) and 15.460 (inboard engines), with cut-off initiated by the firing panel operator. During this firing the combustion chamber pressure at engine 104 appeared very erratic due to a bad transducer, which was subsequently replaced. The bomb in engine 101 detonated 2.740 seconds after ignition command with a period of instability of 8ms. The bomb in engine 107 detonated 2.888 seconds after ignition command with a period of instability lasting 10ms.

Prior to the third instability test bombs were again placed in the two R&D engines. Simulated flight tests were performed on 12 and 13 February. This firing, (SA-51), started at 1630:01.000 CST on 14 February 1968, lasted 15.320 seconds (outboard engines) and 15.200 seconds (inboard engines), with cut-off initiated by the firing panel operator. The bomb in engine 101 detonated 2.520 seconds after ignition command with a period of instability of 10ms. The bomb in engine 107 detonated 2.666 seconds after ignition command with a period of instability lasting 9ms. The Photocon chamber pressure measurements showed no movement during this test firing due to a wiring deficiency at the blockhouse.

Test SA-52 was the fourth planned combustion instability test. Bombs were again placed at the injector faces of the two R&D engines. A simulated flight test was performed on 20 February. The firing, (SA-52), started at 1640:00.000 CST on 21 February 1968, and was planned to last 15 seconds. However the firing was aborted after only 3.484 seconds (outboard engines) and 3.368 seconds (inboard engines) as a result of an explosion and fire at engine 108 (H-4095). The cut-off was initiated by the static test fire detect harness. Primary objectives of the test were accomplished, as the combustion instabilities at engines 101 and 107 had damped out prior to the cut-off signal. The bomb in engine 101 detonated 2.494 seconds after ignition command with a period of instability of 11ms. The bomb in engine 107 detonated 2.634 seconds after ignition command with a period of instability lasting 3ms.

The investigation into the explosion on engine 108 revealed the following. The LOX seal in engine H-4095 began leaking 193ms after turbo-pump first motion. At approximately the same time as the LOX seal began leaking the gear case pressure and the number one bearing pressure began to rise. After the gear case pressure decreased there was a resurgence at 0.914 seconds following turbo-pump first motion. Approximately 360ms after the start of this second gear case pressure rise, rises were indicated on the number one bearing jet pressure and the number one bearing temperature. The gear case accelerometer amplifiers saturated at this time, indicating a detonation in the turbo-pump assembly.

No connection was found between the induced instabilities at engines 101 and 107 and the turbo-pump failure on engine 108. Following the LOX seal failure on engine 108, during test SA-52, NASA recommendations were made to investigate the two LOX seal configurations with respect to carbon nose chipping and leakage rates during the remaining static firings of the S-IB-11 stage. Prior to the next test firing, SA-53, 4 of the engines (102, 103, 105 and 106) were installed with the old vented lip seal and 4 of the engines (101, 104, 107 and 108) were installed with the new improved bellows seals. Following test SA-53 all 8 LOX turbo-pumps were partially disassembled and the LOX shaft seals inspected for carbon nose chipping and sealing surface wear pattern. Following test SA-53 new bellows seals were installed in engines 102, 103, 105, 106 and 108 in order to gain more knowledge of the bellows seal characteristics.

The fifth planned combustion instability firing was cancelled. Following firing SA-52 engine 104 (H-7102) was re-orificed because of high thrust (208,082 pounds), as was engine 106 (H-4093) which had a thrust of 209,233 pounds. The two R&D engines were removed and replaced by the original flight engines from those locations (H-7092 in position 101 and H-4094 in position 107). One of the R&D engines, H-4067, after removal from the stage underwent a series of LOX pump seal tests and checkout of a triple-element thermocouple in the Power Plant Test Stand at MSFC on 19 March 1968.

In addition, engine 108 (H-4095) that had the leakage, explosion and fire in test SA-52, was removed from the stage on 12 March 1968. It was replaced by engine H-4091, an engine that had last been test fired at Rocketdyne on 11 July 1966. The engine had been received at Michoud on 29 July 1966 where it remained until called up as the S-IB-11 replacement. It was shipped from Michoud on 11 March 1968, arriving at MSFC the next day. The engine was installed in the S-IB-11 stage on 16 March.

With the new engine configuration in place the stage was ready for another two firings, the standard short and full duration burns. Prior to that simulated flight tests were performed on 5 and 8 April 1968. The 35 second short duration firing (SA-53) took place at 1640:00.000 CST on 9 April 1968. The firing duration was 35.420 seconds (outboard engines) and 35.304 seconds (inboard engines). Cut-off was initiated by the firing panel operator.

After the installation of bellows type seals in all engine positions preparations were in place for the final, long duration firing. Prior to that, a simulated flight test was performed on 22 April. The full duration firing (SA-54) was started at 1640:00.000 CST on 23 April 1968. Cut-off was initiated by fuel level sensor 2 in tank F-4. Outboard engine cut-off was by the backup timer. Firing duration was 145.328 seconds for the outboard engines and 142.332 seconds for the inboard engines.

In total the stage underwent 7 static firings. Five engines were in place in the stage for all 7 firings with cumulative firing times between 255.2 and 258.6 seconds. Engine 101 had seen 3 firings with a cumulative time of 212.5 seconds. Engine 107 had also seen 3 firings with a cumulative total of 210.3 seconds. Engine 108 was only attached for the last 2 firings with a cumulative total of 177.7 seconds.

The S-IB-11 stage was removed from the tower on 3 May 1968 and loaded on the barge Palaemon on 6 May. The barge departed from MSFC to Michoud on 11 May 1968, arriving 5 days later. The stage was stored at Michoud without Post-Static Checkout being performed. It was placed in environmental storage at Michoud on 20 June 1969 before being removed on 26 March 1973. It was returned to environmental storage on 10 January 1974, together with spare engine H-7112, before being transferred to the ownership of Mason-Rust on 17 May 1974. Ownership was transferred to Boeing Services International on 1 January 1975. For a time in the late 1970s it was displayed horizontally at MSFC, along with the S-IVB-211 stage.

In July 1979 the S-IB-11 stage was moved from Michoud to the Alabama Welcome Centre on the I-65 at the border between Tennessee and Alabama. Here it was erected vertically with an S-IVB upper stage and Apollo capsule of uncertain heritage. It has remained on public view at this location ever since. According to MSFC notes from 1996, in the current display the H-1 engines are not the flight engines but have been replaced by test engines. These notes also state that the S-IVB is a "boilerplate" and the CM is a test article previously used for parachute drop tests at White Sands.

S-IB-12

Summary

This was the final S-IB stage to be static fired. It was never launched and was scrapped after its engines were removed.

Engines

The initial engine configuration was:

Position 101: H-7100
Position 102: H-7101
Position 103: H-7102 – removed from the stage allocation and replaced by H-7098
Position 104: H-7103
Position 105: H-4096 – removed from the S-IB-12 stage and transferred to the RS-27 program
Position 106: H-4097
Position 107: H-4098
Position 108: H-4099

Spare engine: H-7098 – utilized in place of H-7102
Spare engine: H-4073 – utilized in place of H-4096

The stage static firing engine configuration was:

Position 101: H-7100
Position 102: H-7101
Position 103: H-7098
Position 104: H-7103
Position 105: H-4073
Position 106: H-4097
Position 107: H-4098
Position 108: H-4099

Spare engine: H-7097 (not used, transferred to FWV, transferred to S-IB-10 FWV, transferred to EFL)

All the engines were subsequently removed, transferred to the RS-27 program and not replaced.

Stage manufacturing

Clustering of the S-IB-12 stage began on 17 January 1967 in Michoud. The inboard H-1 engines were origi-

Final engine installation in the S-IB-12 stage at Michoud
(28.9.1967) 6759809

East at MSFC on 6 May and performed the standard static firing program. Engineers first removed all eight H-1 engine LOX pump seals from the stage and replaced them with new bellows-type seals to prevent leakage experienced in earlier tests.

A successful propellant load test took place on 3 July. The short trim firing (SA-55) took place at 1652 CDT on 10 July 1968 with a duration of 35.4 seconds. The full duration firing (SA-56) took place at 1640 CDT on 25 July 1968 with a duration of 145.4 seconds. This was the final Saturn IB stage firing and also the final firing on the Static Test Tower East at MSFC.

The official cumulative firing time for the inboard engines was 178.0 seconds and 181.1 seconds for the outboard engines. The stage was removed from the Static Test Tower East at MSFC on 7 August 1968 and prepared for shipment. The S-IB-12 stage was shipped from MSFC on 7 August 1968, arriving back at Michoud on 12 August. Once at Michoud the stage was placed in storage without Post Static Checkout being performed. Engine H-7097 was assigned a Flight Worthiness Verification engine for this stage on 25 June 1969. The stage entered environmental storage at Michoud on 30 June 1969 and remained there until 20 May 1970. The stage was shipped to MSFC on 17 July 1970 where it entered storage, together with FWV engine H-7097, on 28 August 1970. It was stored in MSFC building 4708. It was removed from storage on

nally installed in the S-IB-12 stage in March 1967 and the outboard engines were installed in September 1967, with the final engine being attached on 28 September. Engine H-7102, allocated to the # 103 position, was transferred to the S-IB-11 stage and engine H-7098 (from the S-IB-11 stage) became its replacement.

Engine 105 (H-4096) exhibited LOX seal cavity contamination during initial checkout on the stage and was removed on 12 July 1967. It was replaced by H-4073 which had been removed from S-IB-7 on 28 December 1966. H-4073 had been recertified and underwent engine level testing at Rocketdyne on 20 January 1967 before arriving at Michoud on 7 August 1967. Post Modification Checkout of this engine was completed on 25 August 1967 and the engine was installed on the S-IB-12 stage as the replacement on 28 August.

Stage testing

The S-IB-12 stage underwent Post Manufacturing Checkout from 20 October 1967 to 3 January 1968 following which the stage was shipped from Michoud on board the barge Palaemon on 23 April 1968, arriving at MSFC on 4 May. The stage was installed in the Static Test Tower

S-IB-12 in center in temporary storage at
Michoud, together with S-IB-8 and S-IB-9 (3.1968) 6864868

20 September 1971 and shipped back to Michoud, departing MSFC on 1 October 1971 and arriving on 7 October 1971. The S-IB-12 stage once again entered environmental storage at Michoud on 22 October 1971 where it stayed, together with the FWV engine, until 12 June 1973. During August 1973 the eight H-1 engines were removed from the stage and all were transferred to the RS-27 program, where components would be used on the RS-27 engines for the Delta launch vehicle.

Engine H-7100 was removed on 3 August 1973; H-7101 was removed on 1 August 1973; H-7098 was removed on 2 August 1973; H-7103 was removed on 3 August 1973; H-4073 was removed on 13 August 1973; H-4097 was removed on 10 August 1973; H-4098 was removed on 1 August 1973 and H-4099 was removed on 2 August 1973.

The S-IB-12 stage without engines attached was finally shipped to the Kennedy Space Centre on 8 May 1974 where it is presumed to have been scrapped.

S-IB-13

Summary

The S-IB-13 stage was the first S-IB stage to be built and not static fired. It was the second stage to have its engines removed and for the structure to be scrapped due to a lack of a mission.

Engines

The initial and final engine configuration was:

Position 101: H-7113
Position 102: H-7114
Position 103: H-7115

Fastening fairing to S-IB-13 propellant tank at Michoud
(1.1969) 6973026

Position 104: H-7116
Position 105: H-4109
Position 106: H-4110
Position 107: H-4111
Position 108: H-4112

Spare engine: H-4107 (not used, FWV with this stage, transferred to FWV with S-IB-11)

All the engines were subsequently removed and not replaced. All the engines were transferred to the RS-27 program except H-7113 and H-4109.

Stage manufacturing

All the H-1 engines were installed in the S-IB-13 stage between 5 March and 25 March 1969.

Stage testing

The Post Manufacturing Checkout was not performed

S-IB-13 spider beam on stage assembly fixture
(1.1969) 6973025

Clustering of S-IB-13 stage (2.1969) 6973447

Clustering of S-IB-13 stage
(2.1969) 6973448

as was usual at Michoud. Instead the stage entered environmental storage at Michoud on 22 July 1969. The stage was removed from storage on 2 June 1970 and shipped to MSFC on 17 July 1970, arriving on 28 July. It travelled together with the S-IB-14 stage. At MSFC the stage was never static fired as there was no identified mission. Instead the S-IB-13 stage entered environmental storage at MSFC, in parallel with the S-IB-14 stage in building 4708, on 28 August 1970, where it stayed until 4 February 1972. Four days later the stage was shipped from MSFC, together with the S-IB-14 stage, arriving at Michoud on 14 February 1972. It was placed in environmental storage at Michoud between 27 March 1972 and 19 January 1973. Immediately afterwards the 8 H-1 engines were removed from the stage and most were transferred to the RS-27 program, where components would be used on the RS-27 engines for the Delta launch vehicle.

Engine H-7113 was removed from the stage on 8 February 1973; H-7114 was removed on 30 January 1973; H-7115 was removed on 6 February 1973; H-7116 was removed on 29 January 1973; H-4109 was removed on 26 January 1973; H-4110 was removed on 7 February 1973; H-4111 was removed on 2 February 1973 and H-4112 was removed on 5 February 1973.

Engine H-7113 was allocated as a flight spare on 14 February 1973 and underwent post modification checkout at engine level on 18 July 1973. The engine was shipped to KSC on 19 April 1974 and entered environmental storage with the S-IB-10 stage on 9 May 1974. Subsequently it was in storage with the S-IB-9 stage on 19 November 1974. It finished its life at KSC.

Engine H-4109 was allocated as a flight spare for the S-IB-10 stage, before performing a similar role for the S-IB-9 stage and eventually ending up at KSC.

The S-IB-13 stage, without engines attached, was presumably scrapped at Michoud.

S-IB-14

Summary

The S-IB-14 stage was the second S-IB stage to be built and not static fired. It was the third stage to have its engines removed and for the structure to be scrapped due to a lack of a mission. It was the final S-IB stage to be built.

Engines

The initial and final engine configuration was:

Position 101: H-7111
Position 102: H-7117
Position 103: H-7118
Position 104: H-7119
Position 105: H-4105

Clustering of S-IB-13 stage
(2.1969) 6973449

S-IB-14 thrust structure assembly at Michoud
(1.1969) 6973027

S-IB-14 lower shroud in assembly area in Michoud
(1.1969) 6973028

**Propellant tanks for S-IB-14 in tank
assembly area at Michoud** (1.1969) 6973031

Position 106: H-4113
Position 107: H-4114
Position 108: H-4115

All the engines were subsequently removed, transferred to the RS-27 program and not replaced.

Stage manufacturing

The 8 H-1 engines were installed in the S-IB-14 stage between 14 April and 24 April 1969.

Stage testing

The S-IB-14 stage did not undergo Post Manufacturing Checkout at Michoud. Instead it entered environmental

Installation of the S-IB-14 central LOX tank at Michoud
(3.1969) 6974933

H-1 engine installation in the S-IB-14 stage at Michoud
(4.1969) 6975275

S-IB-14 tank installation at Michoud
(3.1969) 6973935

storage at Michoud on 29 July 1969, where it stayed until 1 June 1970. It was then shipped to MSFC on 17 July 1970, arriving on 28 July. It travelled together with the S-IB-13 stage. At MSFC it entered environmental

storage, in parallel with the S-IB-13 stage in building 4708, on 28 August 1970, where it stayed until 4 February 1972. On 8 February 1972 the S-IB-13 and S-IB-14 stages were shipped to Michoud, arriving on 14 February. The S-IB-14 stage was placed in environmental storage at Michoud on 8 March 1972, where it remained until 23 April 1973. At this point all the 8 H-1 engines were removed from the stage as there was no prospect of the stage ever being launched. This was also the fate of the 2 previous S-IB stages. The engines were all transferred to the Delta RS-27 program where components would be installed in those engines.

Engine H-7111 was removed from the stage on 30 April 1973; H-7117 was removed on 1 May 1973; H-7118 was removed on 2 May 1973; H-7119 was removed on 3 May 1973; H-4105 was removed on 9 May 1973; H-4113 was removed on 10 May 1973; H-4114 was removed on 8 May 1973 and H-4115 was removed on 9 May 1973.

The S-IB-14 stage, without engines attached, was presumably scrapped at Michoud.

S-IB-15

The stage was never assembled, but the propellant tanks were manufactured and stored at MAF in April 1969.

S-IB-14 assembly in Michoud
(3.1969) 6974931

Storage of parts for S-IB-15 and S-IB-16 stages at Michoud
(3.1969) 6974938

S-IB-16

The stage was never assembled, but the propellant tanks were manufactured and stored at MAF in April 1969.

S-IB-14 during final assembly at Michoud
(3.1969) 6974932

S-IV Battleship

Summary

The S-IV battleship was constructed of steel and was tested at SACTO. The vehicle was used for development testing of RL10A-1 and RL10A-3 engines, PU and hydraulic systems, GSE and facilities equipment.

Engines

The stage had 6 RL10A-1 engines for the first 10 firings and 6 RL10A-3 engines for the remaining 17 firings.

Installation of the steam ejector array on an Alpha test stand at SACTO
(1961)

Stage manufacturing

The S-IV Battleship was built by DAC in Santa Monica. It was transported on an open deck barge from the Port of San Pedro in Los Angeles to Courtland Dock on the Sacramento River, and thence overland to SACTO. The stage was installed in Test Stand 1 at SACTO on 11 December 1961 and propellant loading tests followed. The six Pratt and Whitney RL10A-1 LOX/LH2 engines were installed on the stage between February and July 1962 at SACTO.

Stage testing

The first S-IV Battleship firing took place at 1112 PDT on 17 August 1962 with a duration of 10 seconds. All six engines and the altitude simulation system worked successfully. The gear case pressure on engine # 203 exceeded the range of the instrumentation because of excessive pump seal leakage. As a result it was decided to replace this engine with a new one.

The second firing at 1550 PDT on 7 September 1962 was terminated after 13.6 of the planned 60 seconds. For this test the LOX tank pressurization system was to have used the helium heater; however the LOX propellant valve to the heater did not open. The premature engine shutdown was traced to a diffuser vacuum pressure switch, which malfunctioned because of diffuser vibration.

The third firing at 1506 PDT on 15 September 1962 was manually aborted after 28.3 seconds of the planned 60 seconds. The abort was because of high diffuser water temperatures which exceeded the limit of 165 F.

The fourth firing on 24 September 1962 lasted the planned 60 seconds. The LOX valve to the helium

heater failed for the third test running. The next test, on 25 September, involved a planned 180 second ignition of the helium heater only. No engine firings were planned.

The next firing attempt, on 29 September, was manually aborted by a strip chart observer after 42 seconds of the planned 420 seconds because the ullage pressure was not being maintained adequately. On 1 October a planned 420 second firing was aborted after 7.2 seconds by a strip chart observer when the chamber pressure of engine # 203 fell below redline limits. This was determined to be due to a loose "B" nut on the thrust control reference line. An earlier attempt at this test was

Dr von Braun with the S-IV Battleship at Alpha 1 test stand at SACTO
(1962) 9806977

S-IV Battleship firing at the Alpha 1 test stand at SACTO (1962)

prematurely aborted because a tank pressure switch did not pick up and also because the deflector water valve did not open.

The first successful full duration firing occurred on 4 October 1962 with a firing lasting 420 seconds. The bottom section of all six diffusers exhibited extreme erosion and required replacement.

The next attempt at a 420 second full duration firing at 1702 PST on 30 October 1962 was manually aborted after 70 seconds. A malfunction of a helium heater combustion chamber temperature transducer, which caused the chart to indicate above redline values, required that the heater be cut off. The main engines were inadvertently cut off after 70 seconds of firing.

On 3 November 1962 a successful 448 second firing was run to fuel tank depletion. The final RL10A-1 engine test firing in the Battleship stage occurred on 8 November 1962. The test was planned to be run to LOX depletion but was terminated after 38.5 seconds when a pillbox observer noted a fire, which was first visible at 23 seconds and was caused by a thrust chamber fuel leak on engine # 203. The location of the leak had been

patched previously. After the firing additional leaks were found which would have precluded another RL10A-1 firing within a week. On 10 November a decision was made to terminate the RL10A-1 program and switch to the RL10A-3 program.

Total firing time during the 10 firings with the RL10A-1 engines had been 1,137.6 seconds. On completion of the test the six engines were removed.

Towards the end of 1962 Douglas and Pratt and Whitney were modifying six RL10A-3 engines to use as replacements for the RL10A-1s already tested in the Battleship vehicle. The new engines were installed in the S-IV Battleship stage on TS1. After initial difficulty with gaseous nitrogen-water contamination in the engine cool-down valves, DAC completed helium bubbling tests and turbine spin tests of the engines. The first firing in this configuration took place on TS1 on 26 January 1963 with a duration of 468 seconds to LOX depletion.

On 18 and 19 February 1963 S-IV Battleship turbine spin up tests were unsuccessful due to inadequate purge procedures. However, on 23 February a successful spin up was achieved, and two days later on 25 February the second firing took place but was terminated after only 6.5 seconds. This was due to a hydrogen leak which caused a fire at engine # 204. No damage was reported. Firings continued and the seventeenth and final firing in this configuration took place on 4 May 1963 for a duration of 444 seconds. The cumulative firing time with the RL10A-3 engines was 4,302.5 seconds. The S-IV Battleship program ended on 10 May and the vehicle was removed from TS1 at SACTO. On 13 May a one-engine gimbal test was conducted.

During May DAC dismantled the Battleship vehicle. Six of the RL10A-3 engines on hand for the Battleship vehicle were shipped to MSFC, one for use in checking out a seal problem and five for use in Phase III of the SA-D5 Hydrostatic/Dynamics Test vehicle. DAC retained two engines as spares for the ASV stage. On 21 May the S-IV Battleship tank was shipped from SACTO to MSFC. The tank was transported via Courtland Dock and New Orleans. On 7 July 1963 MSFC received the tank for use in the LH2 slosh test program which took place in December 1963. Following this it is assumed that the tank was scrapped.

S-IV-D5 to S-IV-D9 (covering SA-D5 to SA-D9) – Hydrostatic/Dynamics vehicle

Summary

The vehicle was a production configuration stage used for hydrostatic testing at Santa Monica and dynamic testing at Marshall.

Engines

Six RL10A-3 engines, previously used on the S-IV Battleship stage, were installed.

Stage manufacturing

The S-IV dynamic test stage was manufactured by the Douglas Aircraft Company at its Santa Monica plant. The original vehicle that was designated the dynamics stage had problems with the insulation installation that delayed completion. DAC reworked the hydrostatic test vehicle as its replacement, re-designating the vehicle the Hydrostatic/Dynamics vehicle. The Dynamics vehicle was reallocated as the Dynamics/Facilities vehicle for use in wet tests at Cape Canaveral.

The new dynamic test stage was completed in October 1962 and was first used at Santa Monica for structural testing by water pressure. The common bulkhead was

The S-IV-D5 Hydrostatic/Dynamics vehicle being unloaded from the Promise at MSFC (16.11.1962)

cryogenically tested with LN2. The stage was shipped from California to MSFC in Huntsville in a three week journey. The stage was loaded on board the Victory Ship Smith Builder which departed Los Angeles on 26 October 1962. The stage arrived in New Orleans and was transferred to the barge Promise for the journey to Huntsville, where it arrived on 16 November 1962. The initial designation of the stage was the S-IV-D5. MSFC installed some 20 parts and several modification kits in the stage that were unavailable during assembly at DAC.

The S-IV-D5 Hydrostatic/Dynamics vehicle on display at USSRC (2008)

SA-1 — Launch of the first Saturn rocket, on a sub-orbital test flight on 27 October 1961 (left)

SA-2 — Project Highwater released a mass of water into the upper atmosphere aboard the second Saturn rocket on 25 April 1962 (right)

SA-3 — The third Saturn launch on 16 November 1962 was a repeat of the Project Highwater experiment

SA-4 — The third Saturn I launch, on 28 March 1963, included an engine-out test during the flight

SA-5 — The first Block II Saturn I launch on 29 January 1964 included a live S-IV second stage

SA-6 — The second Block II launch, on 28 May 1964, included the first Boilerplate Apollo capsule

SA-7 — Launch of the third Block II Saturn I, on 18 September 1964, included another Boilerplate capsule and the first programmable computer

SA-8 — The first industry-produced booster, SA-8, was the ninth Saturn vehicle launched, on 25 May 1965. It included the Pegasus B micrometeorite satellite

SA-9 — The final Saturn I booster made by NASA launched the first Pegasus satellite on 16 February 1965

SA-10 — The final Saturn I rocket launched the Pegasus C satellite into orbit on 30 July 1965

SA-201 — The first Saturn IB rocket was used for a CM re-entry test on 26 February 1966

SA-203 — The second Saturn IB launch, on 5 July 1966, was used to examine the movement of liquid hydrogen in the second stage in a zero gravity environment

SA-202 — The third Saturn IB launch, on 25 August 1966 was another CM re-entry test

**SA-204 — The Apollo 5 mission
launched the first lunar module
into space on 22 January 1968**

**SA-205 — On 11 October 1968 the first
manned crew were launched aboard Apollo 7**

SA-206 — The first manned Skylab crew were launched towards the Skylab space station on 25 May 1973

SA-207 — On 28 July 1973 the Skylab 3 mission launched the second space station crew

SA-208 — The final Skylab crew was
launched on 16 November 1973

SA-210 — The final Saturn launch took an
Apollo crew into orbit on 15 July 1975
to a historic US-USSR docking in the Apollo-
Soyuz Test Project

Stage testing

The S-IV-D5 stage was installed atop the S-I-D5 stage in the MSFC dynamic test facility on 26 November 1962, the latter having been erected on 13 November. By 17 December the IU, payload adaptor and payload body had been assembled. Phase I testing was delayed to allow for modifications to the test tower. During January 1963 the IU, payload adaptor and payload body were erected on top of the first two stages, forming the SA-D5 vehicle. Propellant mass in the S-IV-D5 stage was simulated using de-ionized water in the LOX tank and Styrofoam balls in the hydrogen tank.

Phase I dynamic testing of the SA-D5 vehicle began on 8 January and was completed on 7 March. The Phase I testing determined the bending modes in the pitch and yaw directions, torsional modes, resonance and frequency response. During the week of 18 March the SA-D5 vehicle was removed from the test stand and the booster stage was sent to Cape Canaveral.

Meanwhile, the Dynamic Test stand at MSFC was modified for an upper stage dynamic test where the vehicle would be supported by a suspension system. On 1 May 1963 the S-IV-D5 stage, the dummy IU and the dummy Jupiter payload were installed in the test stand. Phase II of the SA-D5 dynamic testing was started on 4 May 1963. On 17 May the Jupiter nose cone was replaced with an Apollo boilerplate model comprising a Command and Service Module (BP-9) and Launch Escape System. Testing in this configuration began on 23 May. Phase II of the dynamic testing was completed successfully on 16 June 1963.

During May 1963 DAC dismantled the Battleship vehicle at SACTO. Six of the RL10A-3 engines on hand for the Battleship vehicle were shipped to MSFC, one for use in checking out a seal problem and five for use in Phase III of the SA-D5 Dynamics Test vehicle. The engine used for checking the seal problem was subsequently also used in the Phase III S-IV-D5 stage dynamic testing.

Between 9 and 11 June MSFC received several of the RL10A-3 engines previously used on the S-IV Battleship. These engines were checked out and installed in the S-IV-D5 stage. Gimbal tests of the installed engines were performed between 16 and 28 June. Three more RL10A-3 Battleship engines were received between 27 June and 3 July. These engines were installed in the S-IV-D5 stage on 8 July. The stage was then re-classified as the S-IV-D6 stage.

The S-IV dynamic test stage was removed from the MSFC dynamic test stand on 18 July 1963. On 25 July the S-I-D6 booster stage was placed in the test stand (following its return from Cape Canaveral), followed by the S-IV-D6 second stage on 30 July. The IU was added on 7 August followed by the CSM and adapter on 12 August, forming the SA-D6 dynamic test vehicle. Phase III testing started on 21 August 1963. After the test program was completed successfully on 4 October the vehicle was removed from the test stand on 22 October. It was determined that testing of the SA-7 configuration would be unnecessary because of the similarity between the SA-6 and SA-7 configurations.

The Moog engine gimbal actuator system was installed on the S-IV-D stage for a series of development tests, and on 27 November 1963, the stage was installed once again in the Dynamic Test Stand at MSFC. Gimbal tests, to understand the effectiveness of the Moog actuator in conjunction with the RL10 engines, were started on 5 December, but were terminated on 17 December due to contamination of the hydraulic system. The actuators were returned to the Astrionics Laboratory on 19 December for servicing.

During January and February 1964 the Dynamic Test Stand at MSFC was prepared for a new program with the S-IV-D9 stage. It was determined that a single series of dynamic tests would cover the SA-9, SA-8 and SA-10 configurations. The IU and Apollo SM and ballasted Pegasus payload stages were received in the Test Laboratory and installed on top of the S-IV-D9 stage in the Dynamic Test Stand during the week of 24 February 1964. The Apollo CM and LES were installed during the week of 2 March. Mechanical and environmental testing of these SA-D9 upper stages began on 13 March 1964 and continued until 4 April. Included in these tests were the meteoroid system load and separation tests.

Preparation of the Dynamic Test Stand for the SA-D9 complete vehicle tests was started on 23 April. The S-I-D9 stage was installed on 29 April and the S-IV-D9 stage was installed on 6 May. Dynamic testing began on 20 May and ran through to 9 July. Disassembly of the vehicle began on 10 July and was completed by 17 July 1964. The S-IV-D9 vehicle was then placed in storage at MSFC in Building 4755.

Ground breaking for the Alabama Space and Rocket Center in Huntsville began in July 1968. The following year the S-IV-D9 stage was moved to the outdoor exhibit area. On 26 June 1969 it was moved to a position near the Astronautics Lab at MSFC. Two days later, on Saturday 28 June at 0500 CDT, the stage was transported along Rideout Road to the museum. A number of power lines had to be disconnected, road signs taken down and some poles moved. The stage was mounted vertically on top of the S-IB-D/F stage with a payload on top. The museum opened its doors the following year and the stage has been in the museum ever since.

S-IV Dynamics/ Facilities vehicle

Summary

The vehicle was a production configuration stage used at Cape Canaveral for propellant loading trials and Saturn I Block II launch site compatibility trials.

Engines

Dummy engines were installed in this stage.

Stage manufacturing

Originally in construction as the dynamics vehicle, because of delays in the installation of insulation, this vehicle was re-designated as the Dynamics/Facilities stage and was used in checking out the launch facilities of Block II Saturn I vehicles at Cape Canaveral.

The stage was in structure sub-assembly in Santa Monica from 8 September 1961 to 23 March 1962. The stage was installed in the Assembly Tower on 23 March 1962 and assembly was completed and the stage removed from the tower on 11 May 1962. Checkout operations were conducted from 28 September 1962 until 14 December 1962.

On 18 January 1963 the stage was shipped from DAC's production plant in Santa Monica, via the Los Angeles Dock at San Pedro, to Cape Canaveral, where it arrived on 1 February. It was to be used in checkout of the LC-37B facility. The S-I-D5 booster, that had been involved in the Block II Phase I vibration testing at MSFC, arrived at Cape Canaveral on 15 April. The booster stage was erected on LC-37B on 18 April, and the following day the S-IV Dynamics/Facilities vehicle was erected on top of the first stage.

Stage testing

The series of wet tests were designed to check out LC-37B equipment involved in propellant loading operations. During the week of 24 April propulsion, calibration and mechanical checks were performed. In the first week of May the first stage RP-1 loading tests, the fuel tank pressurization and liquid nitrogen and LOX line tests were completed. There were problems during the S-IV LOX loading test. The first problem arose during a partial test when the LOX replenish system vaporizer ruptured. Problems occurred a second time during a 100% fill when excessive pressure resulted in a broken duct, some deformation of the fill piping, and a malfunction of the LOX tank capacitance probe. The components were repaired and replaced and the LOX loading operations were completed.

Checkout of the LC-37B ended during June. The S-I-D5 stage departed for Huntsville on board the Palaemon on 1 July, whilst the S-IV stage was readied for a test flight aboard the Pregnant Guppy aircraft. On 1 August 1963 the S-IV Dynamics/Facilities stage was flown from Cape Kennedy to Los Angeles on board the Pregnant Guppy, and then on to MSFC for modification and for use in the stratification test program which was to be conducted at MSFC in Huntsville.

The stratification program was a 6-month series of tests to determine the various temperature strata of hydrogen in the S-IV propellant tank during the different levels of filling and pressurization. At MSFC the stage was modified by DAC personnel for the testing whilst MSFC personnel modified the Power Plant Test Stand (usually used for H-1 engine testing) to support the tests. During September and early October the stage was installed in the Power Plant Test Stand, all facility work was completed, and leak checks and functional checkout of the vehicle and facility were completed. The first trailers of LH2 for use in the tests arrived at MSFC on 9 October 1963.

The first stratification test was conducted on 12 and 13 October, producing good measurements in most areas except for flow meter data. On 21 and 22 October technicians conducted calibration tests on the flow meter to determine its reliability and as a result relocated the vent line flow meter. On 6 and 7 November the second and last LH2 stratification test produced satisfactory results. The vehicle was transferred from the Power Plant Test Stand to Building 4705 at MSFC on 14 November 1963. There are no records of the stage after this date and so it was presumably scrapped.

S-IV All Systems Vehicle (ASV)

Summary

Flight-type vehicle used for static testing of the S-IV stage at SACTO to verify the operation of the propulsion system. Destroyed in a test explosion.

Engines

Position 201: P641814
Position 202: P641815
Position 203: P641817
Position 204: P641818
Position 205: P6418xx
Position 206: P6418xx

Stage manufacturing

The S-IV All Systems stage in the Assembly and Checkout Tower at Santa Monica (1962)

The S-IV All Systems Vehicle (ASV) was a heavily instrumented pathfinder for the flight stages. Unlike the Battleship, the ASV was constructed from a flight-weight structure and included a full propulsion system. It was identical to the first flight vehicle, S-IV-5. The ASV was the first flight-type Saturn stage manufactured by DAC in the Santa Monica plant. Structure sub-assembly manufacture took place between 21 July 1961 and 12 January 1962. On that date the stage entered the Assembly Tower at Santa Monica for stage assembly. When this was completed the stage was removed from

Explosion of the S-IV ASV at the Alpha 1 test stand at SACTO (24.1.1964)

the tower on 13 July 1962 and checkout occurred between 13 December 1962 and 16 January 1963. Checkout included the installation of 40 parts that were unavailable during vehicle assembly. The stage was prepared for shipment without the engines installed. These were later installed at SACTO.

Stage testing

The stage was shipped from Santa Monica, via an ocean journey in an open deck barge from San Pedro to Courtland on 1 February 1963. The stage was transported overland from Courtland to SACTO and was installed in Test Stand 2B in mid February. The first propellant loading test on the ASV was repeatedly post-poned because of the non-availability of parts and equipment such as vacuum pumps, umbilicals and ground support equipment.

On 1 April DAC ended the first ASV propellant loading test almost immediately when a leak occurred in the oxidizer umbilical fill nozzle. During the test a LN2 leak caused some cracking in a test stand structural member. DAC made repairs in preparation for a cryogenic test. This was attempted on 9 April, but resulted in the liquid hydrogen tank collapsing after 255 seconds of tanking. Immediately following the collapse technicians experienced difficulty in re-opening the LH2 fill and drain valve. Internal inspection of the tank revealed cracks in the Fiberglass insulation.

Modifications were made to the ASV stage and on 1 May the first successful loading test was performed. However, there were some problems in this activity. After completion of the fill operation the fuel fill and drain valve failed to close. There were also difficulties in opening the LOX fill and drain valve during de-tanking operations. During pressure cycling the ASV bulk-head suffered from internal pressure. Final inspection revealed about 29 hairline cracks in the internal insulation.

On 14 May DAC conducted another loading and pressure cycling test on the ASV. The propellant utilization probe in the LOX tank malfunctioned and the fuel tank insulation suffered additional cracks. The test data from these series of tests revealed that hydrogen leaked through the cracks in the meridian welds into the bulk-head, causing a highly hazardous mixture of hydrogen and air that had been trapped in the bulkhead during its original manufacture.

On 18 May DAC removed the ASV from TS2B and transferred it to the Evaluation and Development building at SACTO in order to clear Test Stand 2B for flight stage acceptance testing. During inspection, performed in June, a number of new weld cracks were found in the meridian welds of the common bulkhead. These were

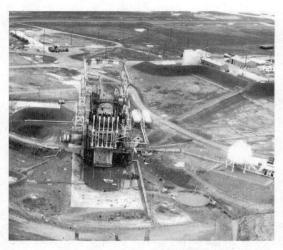

**Damage to the Alpha 1 test stand at
SACTO following the ASV explosion**
(1.1964)

repaired by welding a splice plate over the weld beads and lands of the gore joints. On 6 July 1963 the ASV was installed in TS1 at SACTO.

During July and August 1963 DAC completed buildup for cold-flow testing and pre-test checkout of the stage was also finished. On 21 August 1963 DAC conducted a propellant loading test on the ASV. The test included a functional check of the common bulkhead system, the propellant transfer system, the LH2 tanking system and the fuel tank insulation. Post-test inspection revealed no structural damage to the stage.

Preparations continued towards a first firing in January 1964. Meanwhile the flight stages had overtaken the ASV, having seen the first flight acceptance test in August 1963. The first attempt to fire the ASV occurred on 22 January 1964, but was postponed at 1509 PST due to a LOX umbilical leak. At this point the LH2 tank was empty and the LOX tank was only starting to see liquid. A second attempt, on 23 January, was postponed at 1327 PST when a Hadley fuel fill and drain valve stuck in a partially open position. At this point the LOX tank had been filled and fuel was in the process of being loaded. However, in the terminal phases of the count-down during the third attempt to static fire the ASV stage (countdown 67320), the stage exploded at 1555:39 PST on 24 January 1964.

In the time leading up to the explosion there had been many instances of malfunctioning equipment which complicated and confused the investigation. At 1035 PST on 23 January the primary cold helium shut-off Leonard valve, SN 109, was leaking and was replaced on the following day. However, it was replaced with a valve, SN 123, that had itself been rejected previously, on 7 September 1963. On 23 January the fuel fill and drain Hadley valve malfunctioned because the control

gas was blowing past the actuator piston. The fuel fill and drain valve was removed from the S-IV-6 stage and installed on the ASV stage as a replacement.

Automatic LOX loading was commenced at 1227 PST on 24 January, and completed at 1309 PST. Manual loading of fuel was commenced at 1324 PST and terminated at 1333 PST. Automatic loading was then started and completed at 1348 PST. At 1529 PST all stations were manned and the fuel tank level was at 93%. At 1533 PST the fuel was at 100% and the replenish valve was closed. At 1535 PST the LOX was at 98% and the main LOX fill valve was closed and the replenish valve opened.

Three minutes from the planned firing time (which had been 1555 PST) the "S-IV ready to launch command" was given. At T-15 seconds the deflector plate water was lost. At T-10 seconds cut-off was declared as the diffuser system was lost and there was no steam to generate the altitude conditions needed for firing. The explosion occurred at 1555 PST, 43 seconds after the planned firing time. By 1621 PST the blockhouse had been cleared and observer comments were obtained and securing steps had been outlined.

At the time of the explosion there were 84,244 pounds of liquid oxygen and 16,954 pounds of liquid hydrogen on board the vehicle. As this was the first ground explo-sion involving large quantities of these propellants there was significant interest more in the effect of the explosion rather than the cause. However, the route cause of the problem was that the LOX tank became pressurized well beyond its design pressure. At the time of the explosion the LH2 tank was at 41 psia whilst the LOX tank pressure had risen to an estimated 100 psia. The normal design operating pressure of the LOX tank was 46.5 psia with a vent and relief system in place to prevent the pressure exceeding 48.5 psia. On this occa-sion the relief system failed to operate.

Present day remains of the Alpha 1 test stand at SACTO
(2006)

The review board determined that there were a chain of reasons for the failure of the LOX tank vent and relief valve. The primary cold helium shut-off Leonard valve failed to open. As a result the LOX tank relief valves received excessively cold gas from the helium tanks submerged within the liquid hydrogen tank and stopped relieving. In the days leading up to the failure there were many components that were in short supply including this valve. The day before the static firing the valve on the ASV was not functioning correctly during the checkout. The S-IV-6 flight vehicle was sitting on test stand TS2B and was undergoing cryogenic loading tests at the time. With schedule pressures mounting it was decided to remove the LOX relief valve from the S-IV-6 stage and install it in the ASV. As the valve was on a stage that was still at near-cryogenic temperatures it was decided to pour water onto the valve to heat it up to enable the technicians to remove it. The valve was transferred across, but unfortunately during the warming process moisture was introduced into the valve venting mechanism.

During the ASV countdown an excessively cold helium purge was introduced to all LOX feed lines at the engine interface. The flow of this gas caused the vent valve to freeze closed through water moisture or even oxygen freezing solid in the relief mechanism. As the LOX tank became pressurized the vent valve did not relieve. An observer had been assigned to monitor tank pressure and he was supposed to report any discrepancies to the test conductor during the countdown. At this critical time several issues arose that required attention. The observer was not forceful enough in communicating his concerns to the test conductor who was busy dealing with the several ongoing issues. As a result the pressure continued to rise unabated until rupture occurred.

Detailed analysis of four color films of the incident taken at various distances ranging from 10 to 300 feet identified that the area around the periphery of the common bulkhead had ruptured as a result of the LOX tank overpressure. This had produced a mixing of the bulk propellants and the almost instantaneous explosion. The fireball had reached a maximum diameter of 380 feet and had subsided after 11 seconds.

Two committees of investigation were formed. A NASA committee with DAC membership investigated the cause of the explosion, and first met at Sacramento at 0500 PST on 26 January. Chairman was Dr Kurt Debus, KSC, with Alternate Chairman, Daniel H Driscoll, MSFC. The second investigative committee, under Dr J Gayle, Chief of the Physical Chemistry Section at MSFC, met on 5 February 1964 and investigated the impact of the explosion.

The area around the TS1 was marked out in 50 feet squares and debris collected from each section of the grid was numbered, identified, and its origin and trajectory determined. Damage to surrounding buildings and equipment also was recorded. Using this technique it was possible for the investigators to characterize the size and the force of the explosion. The maximum fragment radius was only 1,500 feet. The results were very significant in that the explosive blast area was far more contained than had been expected at that time from a mixing of LOX and LH2. Damage to the test stand was relatively light; although a neighboring air-conditioned instrument room at the west side of the test stand was demolished by what appeared to be an internal explosion. It was concluded that the substantially instantaneous ignition of the propellants was largely responsible for the relatively low explosive yield. These results had significant implications on the design of future test facilities and launch pads that used these propellants.

One mandatory recommendation of the board was that in the future blast gauges should be installed around test stands in order to directly record the effects of any explosion.

Cleanup of the site commenced on 10 February 1964. To avoid delay and also to avoid the expense of a second ASV, MSFC recommended that the major ASV test objectives could be achieved during the acceptance firings of the S-IV-7 and subsequent S-IV stages.

S-IV-5

Summary

First S-IV stage to be launched on board the first Saturn I Block II vehicle.

Engines

The initial and final engine configuration was:

Position 201: P6418xx
Position 202: P6418xx
Position 203: P6418xx
Position 204: P6418xx
Position 205: P6418xx
Position 206: P6418xx

Stage manufacturing

The meridian welds on the aft half of the common bulkhead were completed in January 1962. Meridian welding of the forward half of the common bulkhead was almost complete when, in April 1962, one of the seg-

Assembly of an S-IV stage at Santa Monica
(Early 1960s)

ments was accidentally damaged. This segment was cut out and replaced with a new one. By June 1962 fabrication of the forward dome, cylindrical tank and thrust structure were complete. The stage was installed in the Assembly Tower at Santa Monica on 17 July 1962.

Assembly of the LOX tank at Douglas Aircraft Company's Santa Monica plant was completed in mid-August 1962, and the stage was moved from the assembly side to the test side of the assembly tower. One month later the stage was moved to the calibration tower for tests and also in October it was returned to assembly for installation of the insulation and cleansing of the tanks.

Pratt and Whitney delivered the 6 RL10 engines to Douglas in October 1962 about a month late because of the lack of checkout equipment. The engines were delivered to Santa Monica for checkout and then to Sacramento for modification. Parts shortages delayed activities during December 1962.

In January 1963 the vehicle was in the final assembly and checkout area where installation of subsystems and components was in progress. In February 1963 continuity and electrical checks were started in areas not affected by the lack of parts. All engines were installed in the stage by 1 March and on 27 March engine gimballing testing was successfully conducted.

DAC completed production acceptance testing of the stage on 8 April and prepared the stage for shipment to SACTO. DAC shipped the stage from Santa Monica on 15 April 1963. Initially the stage travelled by a ship and then barge from the Los Angeles Harbor at San Pedro to Courtland Dock near Sacramento, via Mare Island Naval Shipyard in the San Francisco Bay, where the cargo was transferred to the barge. During the transportation vibration measurements were made on the stage. Severe vibration was recorded for a period of about six hours in high seas. A value of 1.24g was measured against a qualification limit of 1.75g. Following a short overland trucking the stage arrived at

SACTO on 21 April. On 22 April DAC placed the stage in a hangar and began checkout and modifications. On 17 May DAC completed checkout of the stage although many parts were still not installed. The stage was installed in Alpha Test Stand 2B on 22 May. There were still 60 parts missing from the stage at this time.

Stage testing

On 29 May DAC encountered problems during the S-IV-5 simulated flight test that resulted in rework of the guidance signal processor, replacement of the air bearing regulator assemblies and modification of the horizon sensors. Corrective action was taken and system verification accomplished on 12 June during the S-I-5 and S-IU-5 compatibility test. DAC conducted the final simulated flight test on 14 June. On 18 June DAC began pre-static checkout of the S-IV-5 stage. On 27 June DAC conducted the facility checkout, steam blow-down and diffuser water flow tests.

DAC installed redesigned telemetry bracketry on the stage during July. This bracket had been subject to a failed quick-fix after failing a vibration test at Santa Monica. On 11 July 1963 DAC began checkout of the S-IV electrical, telemetry, redline, instrumentation and propulsion subsystems. On 20 July a successful 460 second steam blow-down test was conducted. On 29 July 1963 DAC conducted cold flow turbine spin tests on the S-IV-5 stage. Two attempts were needed to achieve success. In the second run an over-speed of 16,300 rpm was indicated on engine # 203. A series of cold helium bubbling tests were performed on the stage on 31 July.

The first attempt to acceptance fire the stage in Alpha TS-2B at SACTO on 5 August (countdown 67300) was aborted after 63.6 seconds when there were indications of fire in the engine area. Inspections revealed no damage or leaks in the area, and verified that the problem had been due to the threshold temperature of the fire detection system being so low that it was sensitive to the temperature of the steam rising into the engine area. The threshold was changed from 150 F to 200 F.

The second firing (countdown 67301), at 1305 PDT on 12 August 1963, lasted 476.4 seconds until LOX depletion, and was fully acceptable. LOX loading had taken 17 minutes and LH2 loading 28 minutes. Post static checkout was started on 26 August and the stage was removed from the Test Stand 2B on 14 September 1963. The stage was loaded aboard the Pregnant Guppy at Mather Air Force Base, departing on 20 September and arriving at Cape Canaveral on 21 September 1963, where it was transferred to the SAB. Receiving inspection was started on 22 September and only minor discrepancies were found. Weight and balance checks were performed on 8 October.

Installation of an S-IV stage in the Alpha 2B test stand
(Early 1960s)

The stage was mounted atop the S-I-5 stage on Launch Complex 37B on 11 October 1963. Stage checkout was interrupted when contamination was found in the fuel pump inlet valve of engine # 203. Several small pieces of DV1180 polyurethane seal coat were found in the valve. A leak in the facility gaseous hydrogen vent line bellows on 26 November 1963 caused an explosion and ended the first attempt to propellant-load the S-IV-5 stage at LC-37B. At the time about 4,000 pounds of LH2 had been loaded aboard the stage. On 6 December 1963 a propellant loading test of the S-IV-5 stage was performed successfully. The simulated flight test was performed successfully on 22 January 1964 which verified the vehicle readiness for launch.

The SA-5 vehicle was launched at 1124:01 EST on 29 January 1964 from LC-37B. After some 640 seconds of flight earth orbit was achieved by the combination of the S-IV-5 stage, the instrument unit, the payload adapter and the Jupiter nosecone filled with 11,500 pounds of sand ballast. The orbiting payload, designated 1964-05A, re-entered the atmosphere on 30 April 1966 (Universal Time), 29 April 1966 (US time), after 821.40 days in orbit.

S-IV-6

Summary

Second S-IV stage to be launched. First stage to be flown to SACTO.

Engines

The initial and final engine configuration was:

Position 201: P6418xx
Position 202: P6418xx
Position 203: P6418xx
Position 204: P6418xx
Position 205: P6418xx
Position 206: P6418xx

Stage manufacturing

During 1962 DAC began assembling the S-IV-6 stage at

The Alpha site control room
(Early 1960s)

**Installation of the S-IV-6 stage in the
Alpha 2B test stand at SACTO** (30.9.1963)

Alpha test site observation bunker
(Early 1960s)

Santa Monica. Meridian welds of the aft half of the common bulkhead were completed in May 1962. In July welding of the forward and aft sections of the common bulkhead, baffle assembly and tank shell assemblies was completed. The common bulkhead was completed in August along with the thrust structure and aft bulkhead.

Assembly of the LOX tank was completed in September. And by November the vehicle was in the environmental chamber for the installation of insulation. By the end of 1962 DAC had assembled the tanks, installed the insulation, and completed the aft inter-stage compression tests. During early 1963 the stage was in final assembly and mechanical installation was in process. During March and April DAC completed build up of the engines and installed them in the S-IV-6 stage. Parts shortages were hampering completion of component installation. Approximately 30 parts were lacking in April. At the end of April 1963 DAC began the final checkout of the S-IV-6 stage. During May the continuity checks were performed and some rework was required before moving onto checkout. By the end of June the stage was about 3 weeks behind schedule.

DAC completed production checkout of the stage and performed a simulated flight test on 19 July. The flight inverter and other components were damaged and had to be replaced. Following the S-IV-6 checkout DAC decided to complete retrofit and modification of the S-IV-6 stage at Santa Monica rather than SACTO.

Stage testing

DAC completed modifications of the stage and on 27 September the stage was loaded aboard the Pregnant Guppy at Santa Monica Airport. The aircraft landed at Mather Air Force Base later the same day and the S-IV-6 stage was offloaded and transported the short road distance to SACTO. The stage was inspected and installed in Alpha Test Stand 2B on 30 September. During October and November DAC completed pre-static checkout of the stage.

The first propellant loading/turbine spin test of the S-IV-6 was terminated on 19 November because of an umbilical leak with the LOX loaded to the 10% level. The leak was repaired and a successful propellant loading of the stage took place on 20 November. A special Cold Flow Test Evaluation (countdown 67274) was performed in December during which the common bulkhead was purged with GN2 at 3.5 psig for 4 hours. A discovery made from this test was that the common bulkhead had a leak on the LOX side.

The acceptance firing of the stage (countdown 67275) was conducted in Alpha TS-2B at SACTO on 22 November for a duration of 461 seconds. During the test several problems were encountered, and after the test some de-bonding of insulation on the LH2 side of the common bulkhead, between weld seams 5 and 6, was detected. The test problems included failure of the LH2 overfill sensor to function, loss of the engine # 204 hydraulic oil and consequent loss of gimbal control, indication of a leaky check valve on the hydraulic system of engine # 202 and excessive loss of helium control pressure and failure of the LH2 fill and drain valve to open upon command prior to de-tanking. The investigation of the problems, together with the post-static checkout of the stage, was completed on 18 December 1963.

Post-static pressure decay checks between 2 and 17 December 1963 confirmed a leak in the LOX side of the common bulkhead. However as the stage incorporated a pressure relief valve it was concluded not to perform any repairs or modifications. Moog actuators were incorporated into the S-IV-6 stage at SACTO in work that started on 27 December 1963. The stage was removed from the TS-2B on 18 January 1964 after high winds delayed the operation from the previous day. Final checkout of the stage was completed on 17 February in the Evaluation and Checkout Building. The stage left SACTO at 0812 PST on 21 February 1964 and was transported to Mather Air Force Base. It was loaded aboard the Pregnant Guppy aircraft and departed Mather Air Force Base at 1515 PST on 21 February 1964. En route it stopped at Davis-Monthan Air Force Base, Tuscon, Arizona before arriving at Cape Kennedy at 0950 EST on the following day.

Upon arrival at KSC the S-IV-6 stage was moved to Hangar AF at LC-37. Technicians entered the LH2 and LOX tanks to check the cold helium sphere strap tension and for leak checks of the common bulkhead. The leak checks revealed eight transverse cracks that needed repair in the meridian welds on the LOX side of the common bulkhead. The repairs were completed and inspected by 17 March. The S-IV-6 stage was mated with the S-I-6 stage on 19 March. The IU was stacked atop the S-IV stage on 23 March. The Apollo space-craft, BP-13, comprising a CSM, adapter and LES, was placed on top the stack on 2 April 1964.

On 1 April during a routine stage surveillance inspection a depression was noted in the aft inter-stage external skin. It was surmised that a work platform had impacted the vehicle. A doubler was fabricated and bonded over the depression. On 8 May LOX and LH2 loading tests were conducted. During propellant loading tests on 12 May the propellant utilization system malfunctioned due to failure of the LOX sump screen mesh in the S-IV-6 LOX tank. A reinforced screen was added as a repair. The planned launch on 26 May was scrubbed due to an environmental control system compressor malfunction in the ground facilities equipment.

Launch of the SA-6 vehicle occurred at 1207:00 EDT on 28 May 1964.

The inert payload inserted in earth orbit included the expended S-IV-6 stage, the IU and the boilerplate Apollo spacecraft filled with 1,500 pounds of ballast. The payload was given the designation 1964-25A and had a lifetime of 3.31 days, re-entering the atmosphere on 1 June 1964 (Universal time), 31 May (US time).

S-IV-7

Summary

Third S-IV stage to be launched.

Engines

The initial engine configuration was:

Position 201: P6418xx
Position 202: P6418xx
Position 203: P6418xx
Position 204: P6418xx
Position 205: P641853 (subsequently removed and replaced by P641849)
Position 206: P6418xx

Spare engine: P641849 (utilized in place of P641853)

The engine configuration for static test and flight was:

Position 201: P6418xx
Position 202: P6418xx
Position 203: P6418xx
Position 204: P6418xx
Position 205: P641849
Position 206: P6418xx

S-IV stage assembly at Santa Monica
(Early 1960s) 6411809

Stage manufacturing

There were several modifications that distinguished the S-IV-7 stage from S-IV-6. These were the elimination of the LOX tank backup pressurization system and the addition of a non-propulsive venting system on the stage. The backup pressurization system had comprised a number of helium tanks as back up to the helium heater.

The aft and forward sections of the common bulkhead

S-IV stage being lifted into the Alpha 2B test stand
(Early 1960s)

The Alpha site control room
(Early 1960s)

were welded in July 1962 and leak checks were performed in the following month. In October 1962 the common bulkhead and the aft bulkhead were completed. By November the vehicle was in the assembly tower at Santa Monica and the LOX tank was being welded to the LH2 tank. In January 1963 the vehicle was transferred to the environmental chamber for installation of insulation, which was completed in February.

Parts shortages delayed DAC's assembly of the S-IV-7 stage during early 1963 in Santa Monica. By June 1963 the assembly was about 4 weeks late with about 200 parts missing.

In July 1963 the S-IV-7 stage was in final assembly at Santa Monica. Parts shortages continued to hamper assembly. On 27 September installations in the LH2 tank were completed and on 29 September the stage was erected in the vertical checkout stand. The six RL10 engines were installed in October 1963. The engine in position # 205 (P641853) was damaged during gimbal freedom demonstration in the vertical

An S-IV stage being loaded into the Pregnant Guppy
(Early 1960s) 6517512

checkout area on 22 November. The engine was replaced with engine P641849. Final assembly was completed in November and the stage entered final checkout in the vertical checkout tower.

The checkout of the S-IV-7 stage was completed at Santa Monica during January 1964 and the stage was weighed on 30 January. The stage was shipped to SACTO via the Pregnant Guppy on 13 February, flying from Santa Monica Airport to Mather Air Force Base in Sacramento. Because of excessively high surface winds off-loading was delayed overnight.

Stage testing

On 14 February the stage was taken overland to SACTO and placed next to the Alpha Test Stand 2B. The planned erection in the test stand on 15 February was delayed because of a mechanical failure of the test stand crane. Special modifications and repairs were accomplished prior to the installation in the stand, including repainting the stage. Following receipt of repair parts for the test stand crane gearbox on 21 February, the S-IV-7 stage was installed in Alpha TS-2B at SACTO on 22 February 1964 and pre-static checkout of the stage was subsequently successfully completed.

The acceptance firing of the stage was originally scheduled for 16 April. However, due to the replacement of a valve and suspected contamination the firing was rescheduled for 17 April. The firing was again re-scheduled for 21 April to permit inspection of fuel low pressure ducting. The first attempt to static fire the stage, at approximately 1500 PDT on 21 April, failed because a wire omitted in the drawings, and consequently between a relay and the light on the instrument panel, resulted in failure of the panel to indicate "S-IV Ready" at T minus 1 minute 51 seconds in the countdown. On 27 April a marginal electrical umbilical connection resulted in the second attempt to fire the stage to be aborted, after tanking operations.

The stage was static fired successfully (countdown 67336) on 29 April 1964 for a duration of 485 seconds. At manual cut-off both propellants had reached the 0.5% level. Post fire checkout was started on 30 April. On 6 May inspection of the LH2 tank was completed and correction of minor discrepancies was completed prior to removal of the stage from the test stand. The stage was removed from the test stand on 7 May and transferred to the E&D Building for completion of checkout, modifications and preparation for shipment. Post static checks were accomplished successfully and the stage left SACTO by road at 0815 PDT on 9 June 1964. The stage arrived at Mather Air Force Base at 0920 PDT.

The stage was loaded on board the Pregnant Guppy at

Mather Air Force Base but because of poor weather it was decided to delay take-off until the following morning. The aircraft departed Mather at 1020 PDT on 10 June. It landed for refueling at Davis-Monthan Air Force Base, Tucson, Arizona at 1341 MDT on 10 June and departed at 1532 MDT. It refueled a second time at Bergstrm Air Force Base, Austin, Texas, landing at 2020 CDT on 10 June and taking off at 0324 CDT on 12 June. During the landing the outboard wheel of each main gear locked resulting in the blowout of both outboard tires during the landing roll. A third refueling stop was made at Brookley Air Force Base, Mobile, Alabama where the aircraft landed at 0615 CDT on 12 June and departed at 0800 CDT on 12 June. The Pregnant Guppy finally touched down at the Skid Strip, Cape Kennedy, Florida, at 1125 EDT on 12 June. A large number of aircraft failures were encountered during the flights which delayed the delivery of the stage. A visual check of the stage during the unloading at Cape Kennedy revealed that hydraulic fluid had been sprayed on a large area of the forward inter-stage.

The first stage of SA-7, S-I-7 had arrived on 7 June 1964 and was erected on Launch Pad 37B on 9 June 1964. The S-IV weight and balance was conducted on 17 June and the S-IV-7 stage was erected atop the booster stage one day later. The IU was temporarily erected for drill marking on 19 June and finally installed on 22 June. Power was applied to the S-IV stage on 23 June and the spacecraft (CSM BP-15) was erected three days later.

On 4 August the S-I-7 and S-IV-7 stages underwent full pressure checks. The spacecraft LES was erected on 17 August and the All Systems overall vehicle systems test performed two days later. KSC technicians installed on S-IV-7 the newly designed and fabricated non-propulsive venting system received from DAC in August. On 27 August Hurricane Cleo passed the area and the launch complex was secured. On 9 September Hurricane Dora passed the area and once again the launch complex was secured. Fuel loading occurred on 12 September and the Countdown Demonstration Test took place on 14 and 15 September. Booster stage LOX loading and S-IV fuel and oxidizer loading took place on the day of launch.

The S-IV-7 stage formed the second stage of the SA-7 vehicle that was launched from Launch Complex 37B, Cape Kennedy, at 1122:43 EST on 18 September 1964. The stage was given the designation 1964-57A and had an orbital lifetime of 3.86 days before re-entering on 22 September 1964.

S-IV-8

Summary

Launched and survived for 14 years attached to the Pegasus B satellite.

Engines

Position 201: P641855
Position 202: P641856
Position 203: P641863
Position 204: P641860
Position 205: P641861
Position 206: P641862

Stage manufacturing

Assembly of the S-IV-8 stage in Santa Monica was ahead of schedule in early 1963. DAC completed assembly of the LOX tank, installed the stage in the assembly tower and began leak checks and tank calibration.

Assembly of the S-IV-8 stage continued ahead of schedule at Santa Monica during the second half of 1963. In July DAC performed leak checks on the tanks and tank calibration. Installation of insulation was completed in October and the stage was moved from the insulation installation room to the assembly area. The stage remained in the assembly area until DAC received the necessary hardware for installing the stage in the hydrostatic test tower. The stage was installed in the hydrostatic test tower in mid-December 1963 and leak checks were successfully performed. At the end of December the stage was removed from the tower and structural modifications were undertaken.

In January 1964 DAC re-installed the S-IV-8 stage in the hydrostatic test tower at Santa Monica. Leak checks followed by structural modifications were performed. Assembly of the stage in Santa Monica was completed in April 1964.

Stage testing

Checkout of the stage in Santa Monica started on 27 April 1964 and was completed at the end of July. On 7 August DAC shipped the stage to SACTO for static testing. The stage flew on the Pregnant Guppy from Santa Monica Airport to Mather Air Force Base. At the test site technicians positioned the stage horizontally in an area adjacent to Alpha Test Stand 2B. By the middle of August DAC had completed receiving inspection and an ultrasonic inspection of the fuel tank. DAC then initiated installation of the non-propulsive venting systems, electrical modifications and hardware instrumentation installation. The latter included instrumentation for cryogenic calibration.

After removal of the S-IV-9 stage from TS-2B on 27 August test engineers installed the S-IV-8 in the stand the following day. Vehicle and ground support equipment checkout and vehicle modifications continued throughout August and September. From 13 to 26 October DAC performed the S-IV cryogenic calibration weight system tests. These tests involved two fuel and one LOX tank cryogenic calibrations, for the purpose of comparing the stage mass sensor output with the actual mass determined by the weight system. Pre-firing checkouts continued for the next three weeks. Meanwhile DAC replaced the LOX and fuel tank vent and relief valves which malfunctioned during the weight testing. Because of a possible stage/engine control helium oxylube contamination problem the contractor also replaced the 18 engine solenoid valves.

The stage underwent its single static firing in Alpha TS-2B at SACTO on 20 November 1964. The firing duration was 475.8 seconds.

The stage was removed from the test stand on 4 December 1964 and transferred to the Evaluation and Development building.

DAC completed post-static checkout of the S-IV-8 stage in early January 1965 and shipped the stage from SACTO to Cape Kennedy on 23 February 1965, arriving on 26 February. The stage was transported in the Pregnant Guppy aircraft from Mather Air Force Base to Cape Kennedy. The S-I-8 stage was erected on the pad on 2 March 1965, followed by the S-IV-8 stage and S-IU-8 on 17 March. The Apollo BP-26 CM, SM and adapter arrived at KSC on 10 April. The Pegasus B satellite arrived on 15 April and the assembly was erected on the launch vehicle on 28 April. Cryogenic tanking was completed on 11 May, the Flight Readiness Test on 14 May and the CDDT on 20 May.

The SA-8 vehicle, including the S-IV-8 second stage lifted off at 0335:01 EDT on 25 May 1965. In orbit the S-IV-8 stage with Pegasus satellite was given the designation 1965-39A. It re-entered on 3 November 1979 after 5,275 days in orbit.

S-IV-9

Summary

The stage differed from earlier stages by incorporating a blow down vent system in the forward inter-stage. This system augmented the non-propulsive vent sys-

S-IV-9 weight and balance test in hangar AF at Cape Kennedy
(10-11. 1964) 6416600

tem. In orbit for over 13 years together with the Pegasus A satellite attached.

Engines

Position 201: P641857
Position 202: P641850
Position 203: P641851
Position 204: P641852
Position 205: P641836
Position 206: P641854

Stage manufacturing

Welding operations on the forward and aft common bulkhead were completed in August 1962 at the Santa Monica manufacturing plant of The Douglas Aircraft Company. Leak checks of the common bulkhead were performed the following month and the bulkhead assembly was completed in January 1963. Manufacture of the thrust structure and the forward dome were started in December 1962. During February 1963 assembly of the LOX tank was in progress and the thrust structure was attached to the aft bulkhead.

In March 1963 hydrostatic tests were successfully completed and the installation of internal insulation was started. This activity was completed the following month and the stage was painted. Components were installed in the LH2 tank during May 1963. In June bulkhead welds were dye penetrant inspected, but there were delays due to instrumentation shortages. During July and August the stage was in the hydrostatic tower in Santa Monica. The stage was then moved to the assembly area.

By October engine hydraulic and instrumentation build up was in progress and work on the electrical subassembly was almost completed. The six Pratt and

Whitney RL10 engines were received from West Palm Beach during November 1963 and assembly work on the engines was performed. During December 1963 engines 202, 204 and 205 were installed in the S-IV-9 stage, whilst engine 206 was being tested and engine 201 held for parts shortage. These remaining engines were finally installed in the stage during January 1964. Following engine installation the stage was moved into the vertical orientation. DAC completed assembly of the S-IV-9 stage in early February.

Stage checkout in the Vertical Checkout Laboratory (VCL) at Santa Monica started on 10 February 1964 and was completed on 11 April. It was decided not to apply the final paint at DAC as it was not compatible with the thermo-protection paint that would be applied at Cape Canaveral. The stage was removed from the VCL after 5 discrepancies had been corrected.

The stage was weighed on 14 April 1964 in preparation for shipment to Sacramento on 27 April. However, the shipping was delayed due to S-IV-7 stage acceptance firing operations at SACTO. This extra time was put to good use as extra tasks were accomplished at Santa Monica including mounting the stage to the aft interstage.

The S-IV-9 stage departed Santa Monica Airport in the Pregnant Guppy on 8 May 1964 and arrived at Mather Air Force Base at 1907 PDT the same day.

Stage testing

The stage was installed in the Alpha stand 2B later on 8 May. Interface checks and preparations for cryogenic vehicle loading were undertaken during June 1964. The first loading was attempted on 19 June but was terminated because of leakage in the LH2 replenish valve. Three days later the LOX portion of the cryogenic weigh was successfully completed. On 23 June the LH2 cryogenic weigh loading was initiated but was terminated after 1 hour 12 minutes because of indications of gaseous hydrogen in the lower environmental enclosure. However, the test was completed successfully two days later. A second LOX cryogenic weigh loading was successfully performed on 26 June.

During July 1964 there were indications of a substantial amount of oxygen in the core of the common bulkhead which led to an investigation that discovered cracks in the aft face meridional welds of the bulkhead. The leaks were sealed with a repair involving the bonding of doublers over the cracks.

The cryogenic calibration weigh system was used for the first time to calibrate the PU system of the stage. Although problems were encountered, two LOX and two LH2 cryogenic loadings were achieved.

Following the successful cryogenic weighing operations modifications and checkout procedures continued during July. The LOX side of the common bulkhead was inspected and 23 cracks were detected in the meridian welds, three of which allowed leakage. Following repair the bulkhead was leak checked successfully. The LOX tank was then cleaned and flushed, the LH2 LOX feed throughs, LOX sump and LOX low pressure ducts were installed. The helium heater and control helium sphere were installed. Finally the LOX tank was pressure tested at 15 psig for 12 hours to demonstrate that it was leak-tight. During these checkouts and modifications DAC covered the stage with a large plastic bag, circulated warm air around the stage to prevent frost build up and completed a cryogenic propellant weighing program.

The acceptance firing was initially planned for 31 July. However this was rescheduled for 3 August as the workload to prepare the Brown recorders in the blockhouse was more extensive than anticipated. The terminal countdown was initiated at 0912 PDT on 3 August. LOX loading was initiated at 0931 PDT and continued to the 99.75% level. However, the countdown was terminated at 1449 PDT because of numerous difficulties including excessive noise in low level multi-coder system number one. The firing was rescheduled for 6 August.

A successful acceptance firing of 398.94 seconds was performed on 6 August 1964. Cut-off was commanded manually when the diffuser water supply pressure dropped because of failure of two of the facility cooling water pumps. However the duration was considered sufficient to satisfy the data needs. During the firing a malfunction of the fuel tank step pressurization valve occurred, and a rupture occurred in the test stand GH2 collector manifold at the time of the cut-off. Post-firing leak checks of all 6 engines were performed satisfactorily. Dye-penetrant checks of both sides of the common bulkhead were performed and an ultrasonic check of the fuel tank was undertaken.

A visual check of the LH2 tank insulation liner revealed a 30 inch crack which was subsequently repaired. The dye-penetrant checks revealed some cracks that were either burnished out or were repaired with doublers. 72 hour curing of the doubler repair was achieved on 20 August. A successful leak check of the common bulkhead was performed and the tanks were closed out. All tests and checks were completed by 26 August and the S-IV-9 stage was removed from the Alpha 2B test stand at 0800 PDT the following morning. The stage was transferred to the Evaluation and Development Building prior to shipment to KSC.

Because of suspected contamination from fittings incorporating oxylube lubricant it was decided to gas flush the pneumatic systems in order to ensure cleanliness. The RL10 engines were removed and cleaned during this process. DAC also replaced all 18 solenoid valves in the engines.

Shipment to KSC was originally planned for 19 September but slipped due to a delay in the launch date and because of damage to the Pregnant Guppy in Houston, Texas. The extra time was used to fabricate and install an additional fuel tank non-propulsive vent system, authorized by Change Order 413.

Finally, on 20 October 1964 the stage was transported to Mather Air Force Base for shipment to KSC aboard the Pregnant Guppy. The aircraft departed Mather at 0856 PDT the following morning. Some delay was encountered when the aircraft landed at Long Beach for minor maintenance. The stage arrived at KSC at 1630 EDT on 22 October 1964. The stage was offloaded from the Pregnant Guppy the following day and transferred to the SAB. On 18 November 1964 the stage was transferred from the SAB to launch complex 37B. The following day the stage was erected on top of the S-I-9 stage, together with the S-IU-7 stage.

Electrical mating of the three stages took place on 14 December. With the Apollo CM BP-16 already at the Cape, the SM and its adapter arrived from MSFC via the Pregnant Guppy on 13 November. The Pegasus A satellite arrived at KSC on 29 December 1964 after leaving General Electric's Valley Forge, Pennsylvania plant. The Pegasus A was erected on top of the S-IV-9 stage on 13 January 1965 and a day later the Apollo spacecraft was installed.

The All Systems Test ended on 5 February 1965 and the CDDT took place on 12 February. The SA-9 vehicle, with the S-IV-9 upper stage lifted off from LC-37B at 0937:03 EST on 16 February 1965. The S-IV-9 stage, with the Pegasus A satellite attached had an orbital designation of 1965-09A and re-entered on 17 September 1978 after a lifetime of 4,960.65 days.

S-IV-10

Summary

Final S-IV stage to be launched. Survived for 4 years in orbit attached to the Pegasus C satellite.

Engines

The engine configuration for the initial build was:

Position 201: P641864
Position 202: P641870 – subsequently replaced by

Acceptance firing of the S-IV-10 stage in the Alpha 2B test stand at SACTO
(21.1.1965)

P641817
Position 203: P641865
Position 204: P641869
Position 205: P641871 – subsequently replaced by P641884
Position 206: P641886

Spare engine – P641817 (replaced P641870)
Spare engine – P641884 (replaced P641871)
Spare engine – P641885 (not used)
Spare engine – P641887 (not used)

The engine configuration for the static test and for launch was:

Position 201: P641864
Position 202: P641817
Position 203: P641865
Position 204: P641869
Position 205: P641884
Position 206: P641886

Stage manufacturing

During early 1963 DAC in Santa Monica continued fabrication of the S-IV-10 stage. By the end of June DAC had completed fabrication of the forward and common bulkheads and the thrust structure. Structural components for the S-IV-10 stage were all fabricated by September 1963.

In October 1963 DAC installed the stage in the assembly tower in Santa Monica, completed welding and machining operations, and installed the aft inter-stage. DAC moved the stage to the hydrostatic test tower on 4 November 1963 and completed leak checks and tank calibration. On 16 November the stage was moved to the insulation installation room. Following the completion of the installation of insulation the S-IV-10 stage was moved into the hydrostatic tower at Santa Monica

**An S-IV stage being loaded onto the
Pregnant Guppy at Mather Air Force Base**
(Early 1960s)

for leak checks. However, this had been delayed during February 1964 due to parts shortages. DAC completed leak checks and moved the stage from the hydrostatic tower to the assembly area on 15 April. The six RL10 engines were mounted during July 1964. At some stage during testing engine # 202 (P641870) was replaced by engine P641817. Also engine # 205 (P641871) was replaced by P641884.

During July and August 1964 DAC completed final assembly of the last S-IV stage. Final checkout began on 6 August and lasted through to the middle of October. Immediately after a full simulated flight checkout on 14 October preparations began for shipping the stage to SACTO.

Stage testing

The S-IV stage was flown aboard the Pregnant Guppy from Santa Monica Airport, arriving at Mather Air Force Base on 5 November 1964. The next day the stage was moved to the Evaluation and Development building for pre-firing modifications. The major modifications included rework to support installations, instrumentation panel and forward inter-stage bracket; installation of the non-propulsive continuous venting system; rework of the LH2 and LOX point sensors and replacement of the control units as well as rework of the LH2 and LOX overfill control units. It was agreed with NASA that certain parts of the checkout procedure would be eliminated for this stage. Items omitted included cryogenic weighing operations, GSE checkout, vehicle/GSE interface checkout, EMC checkout and instrumentation checks.

It was installed in Alpha Test Stand 2B one month later on 5 December. Pre-firing checkout began on 8 December and continued until 20 January 1965. Countdown 67421 preceded the acceptance firing which was performed on 21 January 1965. The countdown which had

included 4 unscheduled holds totaling 1 hour 12 minutes and 55 seconds was completed in 4 hours 45 minutes and 0.05 seconds. A LOX tank purge was not required for the countdown since the tank had previously been purged with GN2 and a GN2 blanket had been continuously maintained. LOX and LH2 were loaded into the stage tanks. The LOX loading rate was approximately 64% of the design rate. However, the LH2 loading rate was 100% of the design rate.

LOX loading was started at 1039:45 PST when the rapid fill and replenish valves were opened and a LOX main fill line pressure of approximately 70 psia was established. LOX tank pressure continued to rise to 31 psia, beyond the expected limit of 27 psia. For this reason testing was discontinued to allow the pressure to decrease. Loading was continued after 60 seconds. After 2 further short interruptions, LOX loading continued successfully at a rate of 641 gpm, until the 98% mass level was reached, 18 minutes 10 seconds after the start of LOX loading. On-stand leak checks were performed prior to the LOX tank being topped up to the 99.75% level at a rate of 200 gpm.

Fuel loading was started at 1213:45 PST when the fill and drain and replenish valves were opened. The initial fill rate was 524 gpm. At 15% mass level main fill was initiated and the loading rate increased to 2,100 gpm. When the 95% level was reached the fill valve was closed automatically. LH2 replenish continued with normal automatic operation until approximately the 99.25% mass level was reached.

The successful firing lasted for 479.50 seconds on 21 January 1965 and was manually terminated by the cutoff observer when the residual propellants were approximately 0.5%. Post-firing checkout of the stage and securing of the test stand ran until 15 February, after which the stage was removed from Alpha Test Stand 2B on 23 February and stored in the E&D building at

**Present day remains of the Alpha 2
A/B test stand at SACTO** (2006)

SACTO. The S-IV-10 stage was then flown on board the Pregnant Guppy from Mather Air Force Base to Cape Kennedy.

The S-IV-10 stage arrived at KSC on 10 May, the S-I-10 booster stage on 31 May and the S-IU-10 on 1 June. The S-I-10 stage was erected on LC-37B on 2 June, the S-IV-10 on 8 June and the S-IU-10 on the following day.

The Apollo BP-9A SM and SM adapter arrived from MSFC on 21 June, Pegasus C satellite from FHC on 22 June and the BP-9A CM and LES on 29 June. Completed installation of all parts on the launch vehicle was achieved on 8 July. The simulated LOX and LH2 loading tests were performed on 13 July and the Flight Readiness Test on 20 July.

The CDDT was performed successfully on 26 and 27 July. The SA-10 vehicle, including the S-IV-10 second stage lifted off from LC-37B at 0800 EDT on 30 July 1965. In orbit the S-IV-10 stage and Pegasus C satellite was given the designation 1965-60A. It had a lifetime in orbit of 1,465.76 days, re-entering on 4 August 1969.

S-IV-111

DAC began manufacturing of the first operational Saturn I second stage at Santa Monica in December 1962. The common bulkhead was completed as well as fabrication of the structural components of S-IV-111. However, on 30 October 1963 NASA announced its decision to rephrase the manned flights and to delete the Saturn I operational vehicles. This hardware was largely scrapped.

S-IV-112

DAC began manufacturing of the first operational Saturn I second stage at Santa Monica. The common bulkhead of S-IV-112 was completed. However, on 30 October 1963 NASA announced its decision to rephrase the manned flights and to delete the Saturn I operational vehicles. This hardware was largely scrapped.

S-IVB Common Bulkhead Test Article

Summary

Structural test element that failed in test.

The S-IVB Common Bulkhead Test Article in the Alpha I test stand at SACTO (1. 1966)

Stage testing

The S-IVB Common Bulkhead Test Article, which was the modified All-Systems Stage, failed during reverse pressure testing in the Alpha TS1 at Sacramento on 13 January 1966. The Common Bulkhead was severed completely around the circumference at the aft LOX bulkhead joint. It was determined that a design deficiency existed in the joint area. The failed specimen was shipped to Santa Monica for metallurgical and fracture mechanics investigation.

S-IVB-S

Summary

Structural qualification of the Saturn IB second stage and Saturn V third stage achieved via a number of test items.

Stage element activity

The S-IVB-S, structural test stage, was actually a series of structural elements tested and qualified separately. The S-IVB-S was under test at Huntington Beach in the Saturn IB configuration during 1964 and into 1965.

In August 1963 the initial design of the S-IVB-S structural and hydrostatic test stage was completed. Major components for the stage were fabricated by the end of 1963 and in mid-December these components were shipped from Santa Monica to Huntington Beach to await the availability of the assembly tower. On 21 December 1963 DAC began assembling the stage. During January and February 1964 the Huntington Beach assembly tower was used for the first time for assembly of the S-IVB-S stage's LOX tank and cylindrical section. Welding of the forward dome of this assembly in February completed the basic stage structure. This activity had been delayed because of a questionable weld repair in the dome. The decision was made to use the LH2 forward dome built for the S-IVB-D dynamic test stage, as it would provide the desired structural information and integrity. The aft skirt assembly was completed in May 1964 and checkout operations were completed in June 1964.

In the second half of 1964 DAC began testing components to optimize and prove the design load-carrying capability and to establish a margin of safety beyond the maximum expected operational environment. The LOX tank passed the hydrostatic proof test, but during proofing of the LH2 tank on 14 July 1964 the cylinder and forward dome ruptured at approximately limit pressure. X-ray examination showed that failure began with a lack of fusion in one area of the weld seam on the longitudinal portion of the LH2 tank assembly. The LOX tank was returned to Santa Monica for rework in preparation for the thrust structure test. Testing of the forward skirt section continued to the end of December 1964.

The forward skirt assembly was further tested in February and March 1965. In June, DAC completed necessary redesign to correct deficiencies disclosed during tests of the thrust structure. Vibration tests of fuel vent system components, and of the LH2 instrumentation probe structure, were accomplished in July. In July and August test engineers structurally qualified the aft skirt assembly, and on 26 October 1965 completed qualification of the aft inter-stage structure. Completion of the aft inter-stage/aft skirt separation joint test on 11 November left only one structural test, the aft inter-stage retrorocket installation test, to be performed in 1966. After this the stage could be converted for Saturn V configuration tests.

During the first quarter of 1966 the Saturn V S-II/S-IVB Interface Test progressed. Modifications to the S-II Forward Skirt, required to accommodate anticipated acoustic loads, were completed on 29 January 1966. The S-IVB Aft Inter-stage was positioned atop the S-II Forward Skirt in mid-March.

De-stacking of the S-II/S-IVB Interface Test unit was achieved in early July 1966 after the successful completion of tests. The S-IVB/V Aft Skirt specimen was moved to a separate pad for build up for its structural test. The specimen was damaged slightly during preparations and required some rework. The bending moment parameter test was completed on 14 September 1966, with satisfactory results. The maximum ultimate load test was successfully completed on 23 Sep-

tember following a 3 day delay. The elevated temperature parameter test was completed on 4 October. The Main Engine Cutoff Ultimate load test was conducted on 14 October 1966. The failure test of the specimen was conducted on 19 October with satisfactory results. The design limit load was approximately 221% when the dummy aft inter-stage buckled in the vicinity of the maximum compression stringer and the aft skirt tore in the area of the maximum tension stringer.

The S-IVB/V Forward Skirt was subjected to the ultimate test at maximum conditions on 19 July 1966. Failure occurred in bending under approximately 250% design load with axial load held constant at 100% design limit.

The elevated temperature parameter test of the S-II/S-IVB Interface Joint Specimen was successfully concluded on 26 August 1966. On 12 September, during the ultimate axial load/elevated temperature test under MECO conditions, the Interface Joint Specimen failed. Data indicated that the axial load at the time of the failure was about 140.5% of the design limit while the temperature was approximately 520 F.

Testing of the first specimen of the S-II/S-IVB Interface Bolt Tension Test was completed on 11 November 1966.

Structural qualification of the elements making up the S-IVB-S stage were complete by the end of 1966.

S-IVB Battleship (MSFC)

Summary

J-2 Battleship stage that was used at MSFC over many years. It was used as a back up to the SACTO Battleship and for developmental tests of the J-2 engine at MSFC.

Engines

The initial engine configuration was:

Position 301: J-2013

Later configurations included the engines, J-2027, J-2048 and J-2050.

Stage manufacturing

There were two S-IVB battleship stages. The principal battleship for the stage development was tested at SACTO in California. A second battleship was installed in a dedicated test stand at MSFC where it was fired on

The S-IVB Battleship under construction at MSFC
(12.1963)

numerous occasions. This stage was used as a backup for the DAC SACTO battleship and for development tests of the J-2 and J-2S engines at MSFC.

The contractor began preparing the tank assembly site at MSFC in August 1963 and in the week of 21 October began fabrication of the tank. The MSFC battleship stage was nearing the completion of construction near building 4619 at MSFC during December 1963. It was built by the Chicago Bridge and Iron Company, a sub-contractor to DAC. In January 1964 DAC shipped further elements of the battleship tank to MSFC. The battleship, comprising heavy duty tankage together with a flight-type feed system and J-2 engine, was erected at MSFC in February 1964. However, the test stand was not completed until October 1964, following which the battleship was installed in the test stand.

Stage testing

At MSFC, the Test Laboratory proceeded in June and July 1965 to activate the S-IVB test stand for the MSFC battleship. LH2 was loaded into the S-IVB fuel tank on 24 June 1965. A successful engine thrust chamber and start bottle chill test was conducted on 29 June. On 10 July a satisfactory LOX tank loading was conducted. On 13 July an engine spin test was aborted because the power for the re-circulation pumps failed.

On 17 July liquid nitrogen was loaded in the fuel tank to check out the fuel chill down system. A countdown test was performed on 19 July, with LH2 and LOX in

the battleship tanks.

An 8 second test, scheduled for 30 July 1965, was cancelled when the facility heat exchanger thrust chamber chill down coil froze. A successful thrust chamber chill test was conducted on 31 July.

The first successful ignition test for 2.1 seconds was conducted on 2 August 1965. A full duration test lasting 400 seconds followed on 15 September.

The J-2 engine used in the first series of firings (J-2013) was removed from the stage on 4 October 1965 and replaced with up-rated engine J-2027, on 11 October. Also installed at this time was engine gimballing equipment. Four battleship firings using this new engine occurred for long durations ranging from 300 to 432.4 seconds during November and December 1965. All four firings yielded normal results.

The battleship stand at MSFC was used for a total of 117 J-2 and J-2S firings up to 1971.

S-IVB Battleship (SACTO)

Summary

J-2 Battleship stage tested at SACTO and later at Arnold to verify the operation of stage and engine together. Both the S-IVB/IB and S-IVB/V configurations were tested, simulating the Saturn IB and Saturn V launch vehicle configurations respectively.

Engines

The initial engine configuration for countdowns CD 614000-CD 614013 (S-IVB/IB configuration) was:

Position 301: J-2003

The engine configuration for countdowns CD 614014-CD 614032 (S-IVB/IB configuration) was:

Position 301: J-2013

The engine configuration for countdowns CD 614033-CD 614044 (S-IVB/V configuration) was:

Position 301: J-2020

Stage manufacturing

There were two S-IVB battleship stages. The principal battleship for the stage development was tested at SACTO in California. A second battleship was installed

The S-IVB Battleship stage in Test Stand I at SACTO
(1964-1965)

**Final firing of the S-IVB Battleship
stage at SACTO's Beta I stand** (20.8.1965)

in a dedicated test stand at MSFC where it was fired on numerous occasions.

The test program was conducted at the SACTO site of The Douglas Aircraft Company, Sacramento, California from 18 September 1964 to 20 August 1965. The S-IVB static firing program consisted of a series of short and full-duration engine firings to prove the design parameters and to verify the integrity of the stage systems.

The battleship test vehicle comprised the flight stage systems and the battleship tank assembly - a heavy-duty stainless steel, cylindrical vessel with hemispherical heads mounted on a dummy aft inter-stage and the J-2 engine mounted on the thrust structure. The LOX and LH2 tank internal configuration resembled the S-IVB flight stage except for openings provided for special instrumentation, cameras, lighting, and emergency LOX drain provisions.

DAC awarded the S-IVB battleship stage tankage contract to the Chicago Bridge and Iron Company in December 1962. During January and February 1963 the contractor proceeded with fabrication of components. The contractor assembled the tank at SACTO and DAC insulated it. On 2 August 1963 DAC initiated hydrostatic tests and calibration of the S-IVB tank. The tank was then placed horizontally and on 27 September DAC began installation of the internal insulation. In October and November DAC completed S-IVB battleship tank gap filling, seal coating and sanding operations, and began the fiberglass installation.

Stage testing

The battleship tank was installed on the Beta Complex test stand No. 1 at SACTO on 18 December 1963 and engineers began battleship buildup and checkout activities including test stand, Test Control Center, and facility equipment installations and checkout. Safety tests of helium spheres ended successfully at SACTO in March 1964. Battleship test operations started in April with integrated checkout of the electrical systems and pneumatic consoles. Also in April the J-2 engine was delivered from Rocketdyne.

The J-2 engine (J-2003) was delivered to DAC at SACTO on 30 April 1964 and installed on the battleship tank on 4 June 1964. By mid September, checkout of the battleship, GSE and support systems had been completed. Prior to static firing, a number of checks were successfully completed. These included leak checks and complete functional tests of pneumatic, propellant, aft environmental control, and electrical power distribution and sequencer systems.

S-IVB/IB battleship configuration tests took place between 18 September 1964 and 14 May 1965 whilst the S-IVB/V battleship configuration testing was performed between 19 June and 20 August 1965.

Cold flow and chill down testing consisted of a series of non-firing tests, performed in order to establish and evaluate operating procedures for propellant loading, engine purging, venting and a chill down sequence for proper engine start.

Four countdowns, CD 614000, CD 614002, CD 614003 and CD 614004 were required for these tests. The first battleship cryogenic loading with liquid nitrogen and LH2 took place on 18 September 1964. Successful cryogenic loading with LOX and LH2 and the overboard bleed chill-down occurred on 25 September. Successful propellant loading and forward flow recirculation chill-down took place on 2 October, and on 9 October a special thrust chamber chill-down test completed the battleship chill-down program.

The first battleship ignition firing was attempted unsuccessfully on 24 October. On 7 November just after main-stage signal, a failure occurred in the gas generator as a result of overheating during re-circulation chill down. The pressure rise in combustor and LH2 injector manifold destroyed the LOX poppet valve, the number 2 spark plug was blown from its threaded shell, and the LOX injector sense line was burned through and partially consumed. The poppet was blown through the LH2 turbine, where two turbine blades were subsequently destroyed. Corrective measures were taken and the gas generator operated normally during engine operation thereafter.

The first Saturn IB configuration firing occurred on 1 December 1964 and lasted for 10.67 seconds, and the first full duration test came before the end of that month, with a 414.6 second firing on 23 December 1964. J-2 engine J-2003 was replaced with J-2013 on 28 January 1965. The IB firing program with a flight-type engine continued early in 1965 with satisfactory full-duration firings on 31 March, 15 April and 4 May 1965. The Saturn IB series of battleship tests at SACTO ended on 14 May 1965, with an aft inter-stage environmental conditioning test.

Conversion of the stage to the Saturn V configuration began in May and was completed early the following month. This included replacing the J-2 engine with J-2020. The modification also included the installation of 10 ambient helium bottles to the thrust structure for LOX and LH2 tank re-pressurization.

Test personnel attempted the first Saturn V development firing on 19 June, but the test ended with an automatic cutoff after 9 seconds. The second firing on 26 June consisted of a 167 second "first burn" firing followed by a 4 second restart firing which indicated good performance of all systems. During the third firing on 1 July an explosion occurred in the thrust cone area after 2 seconds of the second firing. This caused a fire that damaged wiring and instrumentation.

On 13 August fire interrupted another firing after 16 seconds, but damage was minor. The first two-burn full-duration firing of the battleship occurred on 17 August; firings were for 170 and 319 seconds with a 92 minute simulated coast period between. The final two-burn test came on 20 August for 171 and 360 seconds with a 41 minute coast period.

The S-IVB/V static firing test program consisted of seven firings - two full duration and five short duration, whilst the S-IVB/IB test program consisted of ten firings – four full duration and six short duration.

Workmen removed the battleship stage from the Beta I test stand at SACTO on 3 September 1965 in readiness for transporting the stage for further testing.

In early January 1966 the SACTO S-IVB battleship was transferred to Tennessee. On 8 January 1966 the battleship arrived by barge at South Pittsburg, Tennessee, en route to Arnold Engineering Development Center (AEDC) in Tullahoma for a series of J-2 engine altitude firings.

The series of J-2 and J-2S development firings in the Arnold J-4 Test Cell, some utilizing the battleship stage, ran from July 1966 to October 1968.

Examples of tests run include those during January 1968, when personnel at AEDC conducted several J-2 engine tests. These included 8 low-fuel S-II net positive suction head (NPSH) hot firings and 3 blow-down tests on the S-IVB battleship. The tests also included 6 S-IVB battleship firings to investigate 8-minute restart runs of the 230,000 lb thrust engine.

At the conclusion of testing, at the end of 1968, it is assumed that the battleship stage was scrapped.

S-IVB-F (S-IVB-200F/500F)

Summary

Facilities checkout stage used to check test stand facilities at SACTO and launch facilities at KSC. Part of Saturn IB and Saturn V Facilities vehicles used to check the respective launch site interfaces. Finally converted for use as the Skylab OWS dynamic test vehicle.

Engines

The stage had no engine installed.

Stage manufacturing

The S-IVB-F (or S-IVB-200F/500F), the facilities

checkout stage for the Saturn V vehicle, was manufactured at Douglas' plants during 1964. Fabrication of hardware began at Santa Monica in February 1964. DAC completed all cylindrical tank panels during April. By the end of June the aft LOX dome and common bulkhead were joined, and the forward LH2 dome was ready.

During the second half of 1964 the All-Systems stage, S-IVB-T was cancelled and hardware re-allocated to the S-IVB-F vehicle.

DAC completed the thrust structure assembly and machined the attach ring in October. In late December 1964 DAC moved the stage to Assembly Tower # 2 and joined the forward skirt, aft skirt and thrust structure to the tank section. No engine was installed in the stage. The stage left the Seal Beach Naval Dock on 12 February 1965, aboard the barge Orion, and arrived at SACTO on 17 February, after having been towed up the Sacramento River to Courtland dock. At SACTO, on 18 February 1965, workmen installed the stage in the Beta III Test Stand for use in qualifying the stand for S-IVB-201 acceptance testing.

Stage testing

Propellant loading in manual mode was accomplished without problems on 21 April 1965. The stage and facility checkout in Beta III ended on 1 May with a successful automatic propellant loading test. The stage was removed from the stand on 3 May and underwent post-test inspection of tanks, insulation and welds. No discrepancies were found.

The S-IVB-F stage departed SACTO on 10 June 1965 arrived at KSC on 30 June 1965. Initially, it traveled from SACTO, via Courtland dock, down the Sacramento River to Seal Beach on the Orion, arriving on 13 June. At Seal Beach it was transferred to the AKD Point Barrow for the remainder of the voyage to KSC. Joining it for this part of the journey was the S-II stage simulator. It passed through the Panama Canal on 22 June, calling in at MAF on 26 June, where the S-II stage simulator was offloaded. It was planned to use the stage for checkout of the LC-34, -37 and –39 facilities, although ultimately was only used on LC-34 and –39.

Vehicle erection on Pad 34 began in August 1965. It had been planned to use the S-IB-D/F stage as the first stage of the facilities verification vehicle. However, that stage was damaged during dynamic testing and it was agreed to use the first flight stage, S-IB-1 for facility testing. The S-IB-1 stage was erected on the pad on 18 August. The S-IB-1 served as a spacer for the S-IVB-200F/500F during propellant tankings to verify the LOX and LH2 loading systems. The S-IU-200F/500F was erected atop the S-IVB-200F/500F.

Vehicle checkout of LC-34 began on 18 August and, except for several days lost to Hurricane Betsy, progressed extremely well. Technicians completed the automatic computer controlled propellant loading of the S-IVB-200F/500F on 23 September 1965. On completion of the tests on 29 September, KSC technicians began dismantling the vehicle from the pad. DAC personnel immediately began the process of converting the S-IVB-F stage to the Saturn V configuration.

Conversion of the S-IVB-F, facilities checkout stage, to the final Saturn V configuration was accomplished at KSC in the Low Bay of the VAB. The stage was signed off and available for LC-39 checkout on 25 March 1966.

On 25 March the S-II-F stage was mated with the S-IC-F stage in the VAB. Between 28 and 29 March 1966 technicians at KSC stacked the S-IVB-F atop the S-IC-F and S-II-F. The following day the S-IU-500F was also added to the rocket.

Following completion of the AS-500F, facility checkout vehicle, power was first applied on 13 May 1966. Systems tests were completed on 24 May 1966 and the complete vehicle was rolled out of the VAB towards Launch Complex 39A on the following day. The vehicle travelled on crawler-transporter No. 1 and the journey took most of the day.

On 8 June 1966 AS-500F processing and test activities at LC-39A were interrupted because of the approach of Hurricane Alma. The vehicle was rolled back into the VAB as a precaution. Two days later, with the threat of the hurricane past, the AS-500-F vehicle was again moved out to pad 39A. The journey took about eight hours.

Power and control switching tests were performed during July. Planned tanking tests in August were delayed by the failure of the LOX supply system. Subsequent to repairs to the system, S-IVB-F manual LOX and LH2 loading tests were accomplished on 28 September. During the fast fill, loading was stopped at 52% because of a leak in the swing arm # 6 umbilical connection. It was determined that it was not necessary to rerun the test as all major objectives had been achieved.

By 12 October 1966 the AS-500F vehicle automatic LOX and LH2 loading was satisfactorily accomplished at LC-39A following an aborted attempt on 8 October.

During the S-IVB-F automatic replenish, a leak in the 18 inch GH2 vehicle vent line on the Mobile Launcher developed into a fire. The fire was extinguished by closing the vehicle LH2 vents and applying a GN2 purge of the vehicle venting. The fire caused no damage to the vehicle or the facility. The leak was caused by a ruptured flex bellows in the line.

The S-IVB-F thrust chamber chill down and the terminal count were not accomplished due to LH2 leak and fire.

The launch vehicle drain was satisfactorily accomplished, the sequence being, LOX drain preparations, simultaneous manual S-IVB LH2 and S-II LH2 drain, simultaneous automatic S-IVB and S-II LOX drain and S-IC LOX drain. At this point, AS-500F–1 wet tests were considered complete.

Following completion of testing at the pad, AS-500F was rolled back to the VAB from LC-39A on 14 October 1966. A minor bearing overheating problem on the Crawler Transporter was encountered during the move.

Beginning on 15 October the de-erection of AS-500F took place in the VAB. The LES became detached during a semi-unofficial manual structural test on the Saturn V vehicle in which ropes were attached to the rocket and pulled by workmen on one side whilst other workmen lay on their backs and pushed the vehicle from the other side. The test was to determine the resonant modes of the rocket but resulted in the LES being pulled off. On 15 October the CSM and IU were removed, followed by the S-IVB-F third stage and S-II-F second stage on 16 October, and finally concluding with the S-IC-F first stage on 21 October 1966. The S-IVB-F was placed in storage at KSC.

In early 1967 the S-IVB-F's APS system module was subjected to inert fuel and oxidizer loadings at LC-39A on the Mobile Service Structure. These two inert loadings were followed by hot fuel loading, which was accomplished with no significant problems.

Douglas were directed to ship the stage to MSFC for use as a mockup during the AS-204 Apollo launch. A dummy J-2 engine was installed and the stage de-erected from Low Bay Cell # 2 and prepared for shipment. Prior to the scheduled ship date, the AS-204 mission was scrubbed and shipment of the S-IVB-F stage was cancelled. The stage was returned to the checkout cell at KSC and placed in storage. Early in 1969 it was, in fact, shipped to MSFC.

On 2 January 1970 MSFC shipped the S-IVB-F stage to the McDonnell Douglas plant at Huntington Beach for modification. The S-IVB stage traveled from MSFC Redstone airfield to Los Alamitos Naval Air Station aboard the Super Guppy aircraft. The stage was to be converted into a Skylab Workshop dynamic test article. On 4 December 1970, after conversion was completed, it was shipped from McDonnell Douglas at Huntington Beach to Michoud, aboard the Point Barrow, together with the S-IVB-512 stage. The S-IVB-F stage was

offloaded at MAF and was transported to MSC in Houston, Texas for Skylab Workshop Dynamic testing. The journey was made, leaving MAF on 31 December 1970, onboard the Orion, and arriving at the Clear Lake dock near MSC on 5 January 1971. It was offloaded on 7 January and moved to the MSC acoustic test facility for a series of tests starting on 20 January 1971.

At MSC it underwent a series of tests to verify its bending and vibration characteristics. It was subjected to the lift-off acoustic environment for 15 seconds to qualify the OWS structural design. Acoustic testing was completed on 12 February 1971.

Following the completion of Phase I of the vibro-acoustic test program the S-IVB-F, OWS dynamic test stage, was shipped from MSC on 23 May 1971, onboard the Orion, to MSFC, where it arrived on 4 June 1971 for Skylab Workshop static testing. In June 1974 it was shipped to KSC where it was probably scrapped.

S-IVB-D

Summary

Dynamic test stage used for testing the Saturn IB and Saturn V configurations.

Engines

The engine configuration for dynamic testing of the stage was:

Position 301: J-2006

The engine configuration in the S-IVB-D stage displayed at the US Space and Rocket Center, Huntsville is:

Position 301: J-204

Stage manufacturing

The S-IVB-D Dynamic Test stage was built by Douglas Aircraft Company at Huntington Beach. Fabrication started in August 1963. By the end of December DAC had completed fabrication of the common bulkhead dome and the LOX bulkhead dome and had completed milling the LH2 cylinder skins for the stage. The cylinder skins were formed at the DAC Long Beach facility. During January 1964 DAC completed the LOX tank assembly and delivered it to the Huntington Beach assembly tower. In February the cylindrical tank section was completed. The LOX tank section and the cylindrical section were welded together during March,

**The S-IVB-D stage being hoisted
into the Dynamic Test Stand at MSFC**
(18.1.1965) 6517991

The S-IVB-D stage in the Davidson Center at USSRC
(2008)

and the forward LH2 dome was welded in place, completing structural assembly of the tankage.

DAC also assembled the stage's thrust structure during March. In April the stage was moved into the hydrostatic test tower. During testing on 14 April the LOX tank forward dome wrinkled and had to be repaired and cleaned. Successful proof testing of the stage ended in May.

During August 1964 DAC completed insulation of the S-IVB-D LH2 tank, cleaned the tank, and positioned it in Assembly Tower # 2 where bonding of mounting clips in the tunnel area progressed. In August Rocketdyne delivered J-2 engine J-2006 to DAC for the S-

The S-IVB-D stage in the Davidson Center at USSRC
(2008)

IVB-D. The stage was placed in the Vertical Checkout Tower # 5 at the end of September. This was for installation of the simulated engine and hookup of the hydraulic system. Checkout of the stage was initiated on 13 October and completed on 28 October. Buckling of the LH2 bulkhead necessitated an additional proof

pressure test on 8 November.

DAC then painted the stage, weighed it and attached roll rings before loading it on the States Marine ship Aloha State at Seal Beach on 8 December 1964. The Aloha State sailed on 9 December 1964 for New Orleans via the Panama Canal. On 21 December 1964 the stage was transferred to the river barge Promise. The Promise arrived at MAF the following day where a tug operated a combined tow of the Promise and Palaemon (which had the S-IB-D/F stage on board). The combination was shipped up the Mississippi and Tennessee rivers to MSFC, arriving on 4 January 1965.

Stage testing

The S-IVB-D Dynamic Test stage underwent Saturn IB vehicle testing at MSFC during the first five months of 1965. The Saturn IB flight vehicle had four flight configurations that differ enough for each one to require a separate series of dynamic tests.

SA-201, SA-202, SA-204, SA-205 configuration consisted of the launch vehicle plus the Apollo spacecraft without the LM
SA-203 configuration had no spacecraft and consisted of the launch vehicle and a simple nose shroud
SA-206 configuration incorporated a LM and a boilerplate CSM
SA-207 configuration consisted of the launch vehicle and a complete Block II spacecraft

Dynamic testing was performed in the modified Saturn I Dynamic Test Stand at MSFC. The first stage of the vehicle comprised the Saturn I dynamic test stage (SA-D5) modified to the Saturn IB configuration (S-IB-D/F). The S-IVB-D stage was the upper stage of the vehicle.

Both the S-IB-D/F and the S-IVB-D stages were

installed in the Dynamic Test Stand in January 1965. On 8 February the dynamic test instrument unit, S-IU-200D/500D was installed atop the S-IVB-D stage. The SLA (simulating the nosecone) was airlifted in by helicopter and placed on top of the dynamic test vehicle, although this operation was delayed due to damage sustained by the SLA on landing.

The first phase of testing (SA-203 configuration) started on 18 February 1965 and lasted until 2 March. The second phase of testing (SA-202 configuration) included Boilerplate CSM 27 and started on 15 March. On 27 March the S-IB-D/F spider-beam assembly crossbeam web cracked. This failure necessitated repair of the spider-beam and repeat of some of the SA-202 tests. Dynamic testing resumed on 2 April and continued until 19 April. The third phase of testing (covering the SA-207 configuration) ran from the end of April until 12 May. The fourth and final phase of dynamic testing (SA-206 configuration) ended on 27 May 1965, concluding the Saturn IB total vehicle test program. Vehicle disassembly began immediately in preparation for upper stage tests.

Upper stage testing in the Saturn IB configuration (S-IVB-D, IU, Payload) started on 1 August and continued until 11 September 1965. At this time the S-IVB-D stage was re-designated for Saturn V configuration testing and conversion work took place.

The aft skirt for the S-IVB-D Dynamic stage conversion to the Saturn V configuration departed Seal Beach aboard the AKD Point Barrow on 24 September 1965 and arrived at MSFC on 2 November 1965. This completed stage hardware deliveries to MSFC for this stage.

Saturn V configuration III (S-IVB, S-IU, and Payload) testing in the Saturn IB Dynamic Test Stand was successfully accomplished during the period from 15 October 1965 to 6 November 1965.

Due to the unavailability of the S-II-D stage for configuration I and II dynamic testing a decision was made to use the S-IVB-D stage for Super Guppy flight tests at the start of 1966.

Modifications were made on the S-IVB-D, Dynamic vehicle, to bring the stage to flight configuration for the Super Guppy aircraft flight test. Instrumentation was added to the stage to record aircraft environment. The S-IVB-D was loaded on the Super Guppy at MSFC. No significant problems were encountered during the test flight to Los Alamitos Naval Air Station, near Huntington Beach on 20 March 1966 or the return flight to MSFC four days later.

Data from the flight indicated acceptable conditions for transportation of S-IVB flight stages by Super Guppy air transport.

The S-IVB-D stage was prepared at MSFC for dynamic testing. S-IVB-D Aft Inter-stage stacking began on 23 November 1966 and was completed on 28 November. The S-IVB-D stage stacking operations began on 29 November and were completed on 30 November. The stage was stacked atop the S-IC-D and S-II-F/D stages. The Lower LEM Adapter, LEM and Upper Space LEM Adapter were stacked on 1 December. The CSM and LES were attached on 3 December.

The dynamic test campaign was classified as the AS-500D Configuration I test series. Configuration I was the complete Saturn V vehicle. Testing started in early January 1967 with the roll test being completed on 7 January. However a difference in the hardware configuration compared with the flight vehicle was noticed and it was decided to repeat the test with an improved configuration.

The Configuration I test program included roll testing, completed on 16 January 1967, pitch testing, from 20 to 23 January 1967, yaw testing, completed on 15 February 1967, and longitudinal testing, completed on 26 February 1967. A LOX vent line ruptured during the final longitudinal test. On 6 March MSFC supplied a spare vent line and authorized additional Configuration I tests to verify the Flight Control System.

Testing continued until 11 March 1967 when the final test was performed to verify the flight control system. De-stacking of the AS-500D vehicle began during the second half of March. By 30 March the LES, CSM, IU, S-IVB-D, and S-II-F/D stages had been removed.

Meanwhile the remaining stages i.e. the complete Saturn V minus the first stage were returned to the Dynamic Test Tower for the Configuration II series of dynamic tests that began on 11 May 1967.

Configuration II testing included the yaw test sequence completed on 15 May, the pitch test sequence completed on 2 June, the roll test sequence completed on 10 June and the longitudinal test sequence, started on 13 June and concluded in early July. All programmed testing in the Saturn V Configuration II Dynamic Test Series was completed on 28 July 1967. A one month extension was granted by MSFC to permit rerun of several tests. Following this the stages were separated and the vehicles were removed from the Dynamic Test Tower.

Ground breaking for the Alabama Space and Rocket Center in Huntsville began in July 1968. The following year the S-IVB-D stage was moved to the outdoor exhibit area. On 26 June 1969 it was moved to a position near the Astronautics Lab at MSFC. Two days later, on Saturday 28 June at 0500 CDT, the stage was transported along Rideout Road to the museum. A number of power lines had to be disconnected, road signs taken down and some poles moved. A single R&D J-2 engine, designated J-204, was installed in the stage for display purposes. The museum opened its doors the following year and the stage has been in the museum ever since. On 15 July 1987 the MSFC Saturn V was designated as a National Historic Landmark. After several years of effort, in 2007 restoration work on the stage at the USSRC was completed and it was moved a short distance to the new covered building being constructed to house the complete SA-500D Saturn V. This building, The Davidson Center, held its first function on 31 January 2008 and opened its doors to the public in February 2008.

S-IVB-T

Summary

All-Systems stage cancelled at an early phase during manufacturing.

Stage manufacturing

The S-IVB-T, All-Systems Test stage was manufactured by DAC at Huntington Beach. By the end of December 1963 DAC had milled all the dome segments, constructed the common bulkhead aft dome, and installed the T-ring for the All-Systems Test Stage. Also DAC had begun welding the T-ring to the common bulkhead forward dome, and had begun welding the LOX dome segments as well as the hydrogen cylindrical tank segments.

In early 1964 DAC was halfway through fitting and bonding of the S-IVB-T common bulkhead. This procedure ended in February, and technicians were ready to join the common bulkhead and the aft dome. By the end of March the LOX tank assembly was complete. In April DAC joined the tank sections to the forward dome assembly and completed manufacture of the forward skirt panels. Hydrostatic proof testing of the tank structure occurred in May. In June the thrust structure and tankage were joined and the stage was moved to the insulation chamber for the start of insulation installation.

During the second half of 1964 the All-Systems stage, S-IVB-T was cancelled and hardware re-allocated to the S-IVB-F and S-IVB-ST vehicles.

S-IVB-201

Summary

First use of fully automatic control of the acceptance test firing. First S-IVB-200 stage to be launched.

Engines

The engine configuration for initial build, stage static firing and launch:

Position 201: J-2015

Stage manufacturing

Manufacturing of the stage began in December 1963 with DAC fabricating the stage in the S-IVB Manufacturing and Sub-Assembly Building at Santa Monica. By February 1964, at Santa Monica, the common bulkhead aft dome had been completed except for the dollar weld. In April DAC delivered the LOX tank to the assembly facility at Huntington Beach. The cylindrical tank section seam welds were then completed. Welding of the LOX tank to the LH2 cylindrical section occurred in May 1964. During a review on 14 May it was agreed between DAC and MSFC that the forward LH2 dome formerly allocated to the S-IVB Facilities Checkout stage be used on the S-IVB-201 stage.

During the last week of May DAC completed assembly of the LOX tank, the LH2 cylinder and the reallocated LH2 forward dome and joined the three components to form the propellant tank assembly. DAC then moved the stage to the assembly tower # 1 and in June completed production proof tests before beginning insulation installation.

**Forming LH2 tank wall segments
in the Verson press at Long Beach**
(Mid 1960s)

Meridian welding of the Common Bulkhead segments
(Mid1960s)

After cancellation of the All-Systems stage, S-IVB-T, DAC shifted the engineering effort required to perfect a test stage for "live" testing to development of the S-IVB-201 stage. Additional resources were applied to the stage production in an effort to recover lost time in the schedule.

In August DAC began pre-fitting the insulation tiles in the S-IVB-201 tanks and also began welding assembly of the basic structure. By the end of October tank assembly insulation had been completed. In November the stage was installed in Assembly Tower # 2 for LOX tank beam and baffle rework. Horizontal placement of the tank assembly occurred in December for LH2 tank installations and clip bonding. By the end of December 1964 DAC had completed assembly of the forward skirt, aft skirt and thrust structure.

In late January 1965 the stage was moved to Tower # 6, Position 11, in Huntington Beach for installation of the J-2012 engine, a non-flight engine, which allowed checkout to proceed until a flight-type engine became

LH2 tank segments in the welding fixture
(Mid1960s)

available. After replacing the stage in Tower # 6, Position 9, checkout operations began on 24 February. Electrical continuity tests continued through 30 March when DAC halted the checkout to allow modifications to the LH2 tank and installation of late parts, including the flight engine, J-2015. Engine swap-over occurred between 7 and 14 April. Final inspection of the stage took place on 15 April and it was prepared for shipment to Sacramento.

On 30 April the stage was shipped on board the barge Orion from the Seal Beach Naval Docks, arriving in Courtland on 5 May 1965. It was unloaded at Courtland and transported overland to SACTO on 6 May and installed in the Beta III test stand on 7 May. Stage modifications, not completed at Huntington Beach, were then resumed. Stage pre-checkout activities continued through to 14 July.

Stage testing

On 17 July 1965 an integrated systems checkout of the stage was completed. A simulated static firing was performed on 22 July and on 27 July the stage cryogenic loading test, under the control of the automatic GSE, was completed. Plans to perform a 10 second firing at the conclusion of the loading test were canceled following test problems.

The first attempt to static fire the S-IVB-201 stage in a full-duration test ended prematurely on 31 July 1965 due to component malfunction in Pneumatic Console "A". Stage propellant loading and the automatic countdown sequence proceeded satisfactorily to the point of shutdown. The stage underwent the full duration firing of 452 seconds on 8 August 1965. The firing was controlled throughout, marking the first utilization of a fully automatic system to perform a complete checkout, propellant loading, and static firing of a flight stage.

DAC completed post firing checkout and formally presented the first S-IVB stage to NASA at the Beta III test site on 31 August 1965. The stage was then transported overland to the Courtland Dock on 3 September, loaded aboard the barge Orion, and shipped to Mare Island Naval Shipyard in the San Francisco Bay, California, where it was transferred to the ocean-freighter Steel Executive for the shipment to Cape Kennedy. The stage arrived at KSC on 19 September 1965.

The APS modules for the S-IVB-201 stage were shipped from SACTO to KSC on 14 September 1965. However, in October, the modules were returned to the Douglas manufacturing plant at Santa Monica to correct deficiencies discovered during APS qualification. The modules were corrected and returned to KSC in November.

**Acceptance firing of the S-IVB-201
stage at the Beta III stand at SACTO**
(8.8.1965)

After arrival of the S-IVB-201 stage at KSC it was unloaded from the Steel Executive and moved to Hangar AF for receiving inspection and modification. On 1 October the S-IVB-201 flight stage was erected on top of the S-IB-1 stage on LC-34, just two days after the S-IVB-F stage had been removed from the S-IB-1 stage on the pad. The S-IU-201 was erected on top of the second stage on 25 October.

Power was applied to the S-IVB-201 stage on 26 October and electrical mating of the entire vehicle ended on 10 November. S-IVB-201 testing revealed an oil leak in the engine gimbal system which required checkout and minor modification.

Erection of the SLA-3 and CSM-009 took place on 26 December 1965. The Flight Readiness Review was completed on 21 January 1966 and the LES was attached three days later. The dry CDDT took place on 8 February and the wet CDDT the following day. The Flight Readiness Test was completed on 12 February and RP-1 tanking operations started on 19 February.

The AS-201 vehicle with the S-IVB-201 second stage was launched successfully from launch complex 34 at 1112:01 EST on 26 February 1966.

S-IVB-202

Summary

Stage was unusually subject to three acceptance firings. Launched from Cape Kennedy.

Engines

The engine configuration for initial build, stage static firing and launch:

Position 201: J-2016

Stage manufacturing

DAC began fabricating S-IVB-202 at Santa Monica in February 1964. In April the aft common bulkhead was completed with the exception of the dollar weld. By early June DAC had completed all welds on the common bulkhead to the LOX tank and the LH2 tank. During the second half of 1964 DAC managed to recover about four months' delay in the schedule. In August fabrication and assembly of the propellant tanks were completed. The tank assembly was placed in the Assembly Tower # 4 in September and in October hydrostatic proof testing, cleaning and leak checks were completed. The leak checks revealed that the attach angle welds needed repair. Following these repairs the installation of insulation was started, a task that was completed by December 1964.

In late January 1965 DAC completed stage tank assembly. Hardware installations were completed in the LOX tank by early March and in the LH2 tank by mid-March. In late March the aft skirt and the thrust structure were joined to the stage tank cylinder and on 5 April the assembly was attached to the forward skirt. The flight engine, J-2016 arrived from Rocketdyne on 13 April, underwent receiving inspection, and was attached to the stage on 28 April.

The first phase of checkout of the S-IVB-202 stage in Tower # 6 at Huntington Beach began on 30 April. Insulation resistance checks, continuity checks, cold plate checks, and fore and aft umbilical checks ended in mid-May. Electrical power was applied to the stage for the first time on 14 May. On 7 June the checkout operation was interrupted for completion of additional hard-

**Arrival of the S-IVB-202 stage at SACTO's
Beta III stand as the S-IVB-201 stage departs**
(2.9.1965)

The Beta complex control room at SACTO
(Mid1960s)

ware changes and manufacturing originally scheduled
to occur after checkout. The checkout resumed on 16
June and during July and early August the stage underwent propellant utilization system calibration, automatic checkout, range safety system checkout, exploding
bridge-wire system checkout and engine alignment
checkout.

Phase 1 of the checkout ended at Huntington Beach in
early August. The stage was loaded on board the barge
Orion at the Seal Beach Naval Dock and on 28 August
the barge departed for Courtland Dock. Upon arrival at
Courtland on 1 September the stage was unloaded and
transported by truck to the SACTO test site. The stage
was installed in the Beta III test stand on the following
day. During September DAC completed additional
electrical and propulsion systems modifications and
installations and completed rework of the LOX and
LH2 access door jamb weld before resuming pre-fire
checkout.

Stage testing

Phase 2 of the stage checkout and preparations for
beginning acceptance firing tests ended with a simulated static firing test on 25 to 26 October.

In the first attempt to static fire the S-IVB-202 stage on
29 October the stage primary batteries proved incom-
patible with the GSE power supplies. Secondary batteries were installed on 1 November and a new countdown
for a second static firing attempt was initiated the following day. The firing on 2 November 1965 lasted only
0.41 seconds because of a component malfunction in
the J-2 engine combustion stability monitoring system.
Corrective actions included replacement of accelerometers in the defective system, cable repair and replacement of components which had exhibited abnormal or
erratic operation.

The S-IVB-202 stage was static fired for 307 seconds

on 9 November 1965. The test was partially successful
as a malfunctioning LH2 mass sensing unit in the PU
subsystem prevented a firing to full duration.

Correction of the problem and a minor fire in the primary aft # 2 battery during the countdown on 19
November delayed another full-duration attempt until
the end of November. Countdown operations began on
30 November and the full duration firing was performed successfully on 1 December 1965 with a duration of 463.8 seconds. Cut-off was initiated automatically. The stage was removed from the Beta III test
stand and transferred to the VCL for post-static checkout and modifications.

Following completion of post-static checkout the stage
left SACTO and was transported by road to Courtland
Dock on 15 January 1966. At Courtland the stage was
transferred to the Orion and taken down river to the
Mare Island Naval Shipyard near San Francisco. At
Mare Island the stage was transferred to the Point Barrow for the remainder of the journey to Cape Kennedy,
where it arrived on 31 January 1966. The Point Barrow
had already picked up the S-IV mock-up stage and the
S-IVB-202 aft inter-stage at Seal Beach prior to loading
the S-IVB-202 stage at Mare Island.

The S-IB-2 stage arrived at KSC on 7 February 1966
and was erected on LC-34 on 4 March. The S-IVB-202
stage was erected on top of the booster stage on 10
March. S-IU-202 was erected the following day. The
SLA-4 and CSM-011 were erected on 2 July.

The Flight Readiness Review took place on 11 August.
The S-IVB-202 stage formed the second stage of the
AS-202 vehicle that was launched from complex 34 at
1315:32 EDT on 25 August 1966.

S-IVB-203

Summary

Stage used for in-orbit zero-g hydrogen experiment followed by pressurization to destruction in order to establish structural margins.

Engines

The engine configuration for initial build, stage static
firing and launch:

Position 201: J-2019

Stage manufacturing

The S-IVB-203 stage had a number of changes incorporated in support of a LH2 orbital experiment. The stage was configured to simulate the Saturn V stage functions including orbital insertion and control of propellants during transition to low-gravity operations; orbital coast and LH2 tank venting during three earth orbits; LH2 tank re-pressurization and partial simulation of orbital re-start.

In July 1964 production of major structural assemblies of the S-IVB-203 stage was in process at Santa Monica. At the same time assembly of the cylindrical tank section for the stage was underway at Huntington Beach. Weld repairs were required to the LOX tank assembly at Santa Monica which delayed shipping of the tank to Huntington Beach until October 1964. DAC completed seam welding of the cylindrical section of the LH2 tanks in late September and began welding the forward and aft rings to the section.

After transfer of the LOX tank assembly and the LH2 dome assembly from Santa Monica to Huntington Beach DAC began welding the sections together. The tank assembly was completed in November 1964 after which the assembly was installed in Assembly Tower # 1 for leak and dye-penetrant checks.

In the first quarter of 1965 DAC began sub-assembly of the S-IVB-203 thrust structure, aft skirt and forward skirt. During April LH2 tank insulation, clip bonding and marking provisions for the LH2 orbital experiment television cameras were completed. Installations in the LH2 and LOX tanks continued during May and June as did modifications for the LH2 orbital experiment.

Engine J-2019 arrived from Rocketdyne on 1 May and entered engine checkout. The aft skirt and the thrust structure were joined to the tank assembly on 26 June 1965. Forward skirt installations were completed and the skirt was attached to the stage on 12 July. The flight engine was attached on 22 July.

Checkout operations were started towards the end of July incorporating some substitute hardware in lieu of late hardware.

With the stage in the # 1 Tower the LOX tank beam and baffle rework was performed. In September 1965 tests of the S-IVB-203 propellant tank assembly revealed structural weaknesses around the manhole cover in the LH2 tank. Repair of cracks solved the issue and factory checkout resumed on 20 September, before completion on 9 October.

Complete vehicle inspections ended on 14 October and shipping preparations were completed on 27 October. The stage was transported to Seal Beach Naval Docks on 29 October 1965 and loaded on the barge Orion for the trip to SACTO. The barge transported the stage to

The S-IVB-203 stage being hoisted into the Beta I stand at SACTO (3.11.1965)

Courtland Docks where it was off-loaded on 1 November 1965 and transported the remaining distance to SACTO overland on a truck.

Most of the modifications needed to ensure that the S-IVB-203 stage simulated a Saturn V stage were completed at SACTO.

Stage testing

The stage was installed in the Beta I test stand at SACTO on 3 November and installations, modifications and limited checkout of the stage progressed through December 1965. Completion of tank close-out was delayed due to engineering changes, rework, repairs and parts shortages.

On 17 January 1966 DAC personnel at SACTO began propulsion subsystem checkout of the S-IVB-203 stage. A test accident damaged the recirculation pump and necessitated its replacement. Pre-static testing was completed in the Beta I test stand on 11 February 1966 with initiation of the simulated static firing test.

Following unsuccessful attempts to static fire the stage on 18 and 22 February, the stage underwent a successful 284.9 second firing in Test Stand Beta I on 26 February 1966. Following the firing there was a simulated two orbit coast period with engine restart condition preparations after each orbit. The stage was removed from the Beta I stand on 19 March in preparation for completion of post-static checkout.

The stage left Mather Air Force Base aboard the Super Guppy on 4 April 1966 bound for Cape Kennedy, where it arrived on 6 April.

The S-IVB-203 stage was erected on top of the S-IB-3 stage on 21 April followed by the S-IU-203 and nose cone on the same day.

Initial power was applied to the S-IVB-203 stage on 28 April and a full pressure test was conducted on 26 May. On 7 June the LOX and LH2 loading test was accomplished. The Flight Readiness Test was performed on 27 June and the CDDT was performed between 29 June and 1 July.

The S-IVB-203 stage formed the second stage of the AS-203 vehicle that was launched from complex 37B at 0953:17 EST on 5 July 1966. Following a zero-g hydrogen experiment with this stage the pressure in the hydrogen tank was allowed to increase until stage destruction occurred 6 hours 20 minutes after launch.

S-IVB-204

Summary

Stage originally stacked on pad 34 for the proposed Apollo 1 flight. After the fire it was launched from pad 37B carrying the first lunar module.

Engines

The engine configuration for initial build, stage static firing and launch:

Position 201: J-2025

Stage manufacturing

During the second half of 1964 DAC began sub-assembly of major structural hardware for the S-IVB-204 stage at Santa Monica. DAC fabricated the forward dome segments and assembly, and the cylinder segments and assembly. However, considerable difficulties were experienced in fabricating the common bulkhead. Replacement of a large section of undersized core at one location was necessary and at another location mis-fit of the forward skin to the core necessitated multiple layers of adhesives. This resulted in a wavy forward skin pattern. Weld deficiencies in the LH2 tank seam welds also required rework at Huntington Beach.

The LOX tank was joined to the cylindrical LH2 tank walls on 28 January 1965. Proof testing of the tank was performed on 5 February 1965. Dye-penetrant and leakage checks of the stage propellant tank were completed on 19 February. LH2 tank insulation installation was started on 25 February and completed on 14 May. Subassembly of the stage aft skirt, forward skirt and thrust structure began in March. LH2 tank installations ended on 25 June and LOX tank installations started.

The J-2025 engine was shipped from Rocketdyne's Canoga Park on 16 July 1965, arriving at Huntington Beach later the same day. Post modification checkout of the engine was completed on 29 September 1965 and the engine was installed in the S-IVB-204 stage the following day.

The thrust structure and aft skirt were joined to the stage in early September followed by the forward skirt on 10 September. Factory checkout of the stage was initiated on 25 September and completed on 15 December 1965 with a simulated flight test. Vertical inspection was completed on 20 December 1965.

Stage testing

Post Manufacturing Checkout of the S-IVB-204 stage was completed on 22 December 1965 and the stage left Huntington Beach on 10 January 1966, arriving at The Douglas Aircraft Company's SACTO test facility on 14 January. It travelled on the barge Orion from Seal Beach Naval Docks to Courtland Dock on the Sacramento River, and thence by road to SACTO. The stage was installed in the Beta III test stand on 15 January 1966 and pre-static checkout was completed on 17 March 1966. During this time the Integrated Systems Check was performed on 10 March and the Simulated Static Firing Test on 14 March. The stage underwent a single firing of 451.2 seconds duration on 18 March 1966 and it was removed from the test stand on 26 March.

Post static checkout was completed in the VCL on 28 April 1966, with a simulated flight test being performed on 25 April. This test was subsequently repeated on 3 May, following which the stage was place in storage at SACTO. The stage departed Mather Air Force Base in the Super Guppy on 6 August 1966, arriving at KSC later the same day.

The stage was erected on Launch Complex 34 on 31 August 1966, on top of the S-IB-4 stage, as part of the planned first manned Apollo flight in February 1967.

S-IVB-205

Summary

Vehicle used as the second stage to launch the first manned Apollo flight, Apollo 7.

Engines

The engine configuration for initial build, stage static firing and launch:

Position 201: J-2033

Stage manufacturing

DAC welders at Huntington Beach began forming segments for the LH2 tank cylinder in March 1965. These were completed in April after which they were x-rayed. Joining of the LOX tank and the LH2 tank shells ended in early May. Welding of the forward dome to the LH2 tank, which began on 20 May, ended in early June. Later in June technicians detected a crack in the jamb weld of the LOX tank during leak and dye-penetrant checks. Subsequent repair operations did not end until late September.

Insulation installation in the LOX and LH2 tanks which began in July ended in November. On 9 December the aft skirt and thrust structure were joined to the tank assembly and the following day the forward skirt was attached. The stage was installed in Checkout Tower # 6 at Huntington Beach on 3 January 1966. Installations, modifications and checkout proceeded.

The J-2033 engine was shipped from Rocketdyne's Canoga Park on 16 August 1965, arriving at Huntington Beach the following day. Post modification checkout of the engine was completed on 29 November 1965 and the engine was installed in the S-IVB-205 stage on 11 January 1966.

Stage testing

Post Manufacturing Checkout of the S-IVB-205 stage was completed on 24 March 1966 and the following day the stage was removed from the Checkout Tower # 6. The stage left Huntington Beach on 8 April 1966. It was transported the short distance to the Seal Beach Naval Docks and loaded aboard the barge Orion, which departed on 9 April. The barge arrived at Courtland on 13 April, from where the S-IVB-205 stage was transported to SACTO by road later the same day. The stage was installed in the Beta III test stand on 15 April 1966 and the Pre-Static Checkout was performed. This was completed on 1 June and the simulated acceptance fir-

The S-IVB-204 stage being hoisted into the Beta III stand at SACTO
(15.1.1966)

The payload, CSM-012, was attached on top of the stages. Power was applied to the stage on 15 September and a full pressure test took place on 28 October. The Integrated Umbilical-In Test took place on 25 January 1967. However, the mission was cancelled when the CM capsule caught fire on 27 January 1967 killing the three astronauts on board. After the Apollo 1 fire the S-IVB-204 stage was de-erected from LC-34 on 3 April 1967.

It was then re-erected, this time on Launch Complex 37B on 10 April 1967, and underwent a requalification test program. The S-IVB-204 was placed on top of the S-IB-4 first stage, which had been erected on the pad three days before. The S-IU-204 was erected on 11 April, followed by the first lunar module, LM-1, enclosed by SLA-7, which was erected on 19 November 1967. Power was applied to the stage on 18 April 1967. The Flight Readiness Test was performed on 23 December 1967 and the CDDT was completed on 20 January 1968, after being terminated the day before.

The S-IVB-204 stage was eventually launched at 1748:08 EST on 22 January 1968 when it became the second stage of the Apollo 5, AS-204/LM-1 vehicle that launched the first lunar module.

The S-IVB-204 stage received the international designation of 1968-07C as it remained in orbit for 0.69 days before re-entering the atmosphere.

**S-IVB-205, S-IVB-206 and S-IVB-207
outside the VCL at SACTO**
(10.1967) 6760458

ing was performed on 23 and 24 May. A single firing of
437.5 seconds duration was performed on 2 June 1966.
Post fire checkout in the Beta III stand took place from
3 to 24 June. The All-Systems Test was run from 24 to
28 June.

Post static checkout was completed on 1 July 1966, following which the stage was removed from the Beta III
test stand on 5 July. It was placed in storage for nearly
two years because of the delays to the Apollo program.
Initially it was stored in the VCL Tower # 2 at SACTO
before being transferred to the VCL Tower # 1 on 7
August 1967 for cleaning prior to modifications. On 28
September 1967 it was placed in the VCL vertical
checkout stand for post-storage modifications. Howev-

**The S-IVB-205 (horizontal), S-IVB-206 (on bird cage) and
S-IVB-207 stages in storage in the VCL at SACTO** (1967)

er, these modifications were not started until 7 March
1968. The stage finally departed Mather Air Force Base
in the Super Guppy on 6 April 1968, arriving at KSC
two days later. It was transferred to the VAB low bay on
8 April.

The stage was erected on top of the S-IB-5 stage at LC-
34 on 16 April. Following this the S-IU-205, SLA-5
and CSM-101 were stacked. The CDDT was conducted
between 12 and 16 September. The Flight Readiness
Test was conducted on 26 and 27 September.

The S-IVB-205 stage was eventually launched at
1102:45 EDT on 11 October 1968 when it became the

**A present day view of the Vehicle
Checkout Laboratory at SACTO**
(2006)

**S-IVB-205 stage being loaded on board
the Super Guppy at Mather Air Force Base**
(6.4.1968) 6866307

second stage of the AS-205 vehicle that launched the first manned Apollo spacecraft, Apollo 7.

The S-IVB-205 stage received the international designation of 1968-89B as it remained in orbit for 6.83 days before re-entering the atmosphere on 18 October 1968.

S-IVB-206

Summary

Stage originally manufactured in 1966 and erected on pad 37B in January 1967. First stage to be flown from Huntington Beach to SACTO. Two static firings needed because of replacement of the J-2 engine LOX turbo-pump. After 5 years of storage the vehicle finally flew in 1973 as part of the first manned Skylab rocket.

Engines

The engine configuration for initial build, stage static firing and launch:

Position 201: J-2046

Stage manufacturing

In September 1965 DAC re-assigned the S-IVB-207 common bulkhead to S-IVB-206 because of TIG weld difficulties in forming the S-IVB-206 bulkhead. The bulkhead was joined to the LOX aft dome and LOX tank in early September. In late September the LOX tank was joined to the LH2 tank cylinder, followed by the attachment of the forward dome to the tank assembly on 7 October. Jamb weld repairs and hydrostatic tests on the tanks ended in November and insulation installations were started. Tank proof testing took place on 8 November 1965. Insulation and bonding were completed on 14 January 1966.

The S-IVB-206 stage being unloaded from the Super Guppy at Mather Air Force Base (14.4.1967)

S-IVB-206 removal from the VCL at SACTO
(10.1967) 6760459

The J-2046 engine was shipped from Rocketdyne's Canoga Park on 20 December 1965, arriving at Huntington Beach the following day. Post modification checkout of the engine was completed on 3 February 1966 and the engine was installed in the S-IVB-206 stage on 31 March 1966.

Stage testing

Post Manufacturing Checkout of the S-IVB-206 stage was completed on 26 May 1966 and the stage left Huntington Beach on 30 June 1966, arriving at The Douglas Aircraft Company's SACTO test facility on the following day, 1 July. The stage had flown in the Super Guppy from Los Alamitos Naval Air Station to Mather Air Force Base. It was installed in the Beta III test stand on 6 July and pre-static checkout was completed on 12 August 1966, following an Integrated Systems Test two days before. A simulated static firing was performed on 11 and 12 August to verify the countdown procedure.

Countdown CD 614069 was initiated on 17 August but was terminated at T-8 hours to allow for the addition of seven accelerometers. Two firings of the stage were performed with a cumulative time of 500.4 seconds. The first firing of duration 433.7 seconds was performed at 1524:37.77 PDT on 19 August 1966 (CD 614070), with engine cut-off being initiated by an observer due to LH2 pump inlet conditions. A routine post firing LOX turbine swab check revealed numerous metallic particles in the turbo-pump. The turbo-pump was replaced, necessitating a short duration J-2 engine re-firing to calibrate the new pump to the J-2 engine.

This second firing of duration 66.6 seconds took place on 14 September 1966 (CD 614072). This second firing successfully verified the performance of the J-2 engine following the replacement of the LOX turbo-pump assembly. Engine cut-off was initiated from the test conductor's console as planned. The stage was transferred to the VCL at SACTO on 4 October 1966. The All-Systems Test took place on 17 October.

Post static checkout was completed on 4 November 1966, following which the stage was shipped from Mather Air Force Base in the Super Guppy on 13 December 1966, arriving at KSC on 14 December 1966. The stage was transported to LC-37B and erected on top of the S-IB-6 stage on 23 January 1967. Testing continued successfully until early March 1967 when the program was re-aligned following the Apollo fire in January. The S-IVB-206 stage was de-erected and moved to Hangar AF where it was prepared for shipment to SACTO.

The stage departed Cape Kennedy aboard the Super Guppy on 13 April 1967 and arrived at Mather Air Force Base the next day in preparation for long term storage at SACTO. On 12 November 1967 the stage was moved from the VCL to the Beta Complex at SACTO for completion of post-storage checkout operations. It was installed in the recently refurbished Beta III test stand on 22 November 1967 for these tests. Post-storage checkout operations were completed on 29 December 1967 and the stage systems were re-verified in a successful AST. The stage was removed from the Beta III test stand on 11 January 1968 and placed in storage in the VCL at SACTO.

The stage was later shipped from SACTO to McDonnell Douglas' Huntington Beach manufacturing plant on 3 August 1970. Finally the stage was flown from Los Alamitos Naval Air Station, near Huntington Beach, to KSC aboard the Super Guppy aircraft. It left Los Angeles on 23 June 1971 and arrived at KSC the following day. The S-IVB-206 stage was placed in long term storage at KSC from 15 October 1971 to 17 April 1972. A propulsion subsystem checkout was completed on 21 August 1972.

The CSM-116 arrived at KSC on 19 July 1972. The S-IB-6 stage and the S-IU-206 both arrived at KSC on 22 August. The S-IB-6 stage was stacked atop the milk stool on ML-1 in High Bay 1 of the VAB on 31 August. The S-IVB-206 stage was stacked on top on 5 September, followed by the IU two days later. This marked the re-uniting of these first and second stages after a 6-year period. A dummy CSM (BP-30) was added together with a stub section of the LES on 8 September, and the stack was rolled out from the VAB to pad 39B on 9 January 1973.

A propellant loading and All-Systems Test was completed on 30 January 1973. After a series of fit checks the stack was rolled back to the VAB on 2 February. On 9 February 1973 the CSM-116 was mated to SLA-6A and on 20 February the assembly was moved from the O&C building to the VAB. On 20 February BP-30 was de-stacked followed by the installation of CSM-116 on the following day. The BP-30 CM ultimately found a home at the top end of the restored Saturn V in the Saturn V center at KSC after temporarily being located in Fort Worth.

The LES was installed on 24 February before the rocket's final roll back to pad 39B on 26 February 1973. The Flight Readiness Test was conducted on 5 April. RP-1 fuel was loaded on 23 April, and the Countdown Demonstration Test (wet) took place on 3 May 1973. The dry CDDT took place the following day. The rocket was due to be launched on 15 May but the launch was delayed following the problems experienced by the Skylab space station during its launch. The S-IVB-206 stage formed the second stage of the AS-206 vehicle which launched the first manned crew to the Skylab space station at 0900:00 EDT on 25 May 1973 on board Skylab 2.

The S-IVB-206 stage received the international designation of 1973-32B as it remained in orbit for less than half a day before re-entering the atmosphere.

Current view inside the VCL at SACTO (2006)

S-IVB-207

Summary

Manufactured in 1966/67 and placed in storage until 1972. Part of the vehicle that launched the second manned Skylab crew.

Engines

The engine configuration for initial build, stage static firing and launch:

Position 201: J-2056

Stage manufacturing

In September 1965 DAC re-assigned the S-IVB-207 common bulkhead to S-IVB-206 because of TIG weld difficulties in forming the S-IVB-206 bulkhead. Jamb weld rework was conducted on the S-IVB-207 forward and aft domes in December 1965. Additional tests following repair operations revealed further defects which subsequent tests indicated would not degrade stage integrity. By the end of December 1965 joining of the LOX tank to the LH2 tank was completed. This was followed by welding of the forward dome to the tank assembly. Proof testing of the tank assembly took place on 14 January 1966. Insulation and bonding was completed on 11 March 1966.

The J-2056 engine was shipped from Rocketdyne's Canoga Park on 25 February 1966, arriving at Huntington Beach later the same day. Post modification checkout of the engine was completed on 28 April 1966 and the engine was installed in the S-IVB-207 stage on 11 May 1966. The following day factory checkout of the stage was started in Huntington Beach.

Stage testing

Post Manufacturing Checkout of the S-IVB-207 stage was completed on 21 July 1966 and the stage left Huntington Beach on 30 August 1966, arriving at The Douglas Aircraft Company's SACTO test facility on the following day. The stage had flown in the Super Guppy from Los Alamitos Naval Air Station to Mather Air Force Base. The stage was installed in the Beta I test stand on 1 September where pre-static checkout was completed on 18 October 1966, the integrated systems test having been completed on 5 October. The simulated static firing was satisfactorily performed on 11 and 12 October to verify the countdown procedure.

At the start of propellant loading there was a severe leak in the LOX sled main fill valve. The valve was left open for the duration of the countdown with main LOX flow being controlled through the facility transfer line valve. The LH2 loading was completed in 31 minutes 55 seconds. The stage underwent the single acceptance firing of 445.6 seconds duration successfully at 1118:45.863 PDT on 19 October 1966 (CD 614074). Engine cut-off was initiated through the PU processor when LOX was depleted below the 1% level. The stage was transferred to the VCL at SACTO on 3 November 1966, and an All-Systems Test took place on 18 November.

Post static checkout was completed on 30 November 1966, following which the stage was placed in long term storage at SACTO on 19 January 1967. It was eventually transported from SACTO to the Seal Beach plant of North American Aviation on 1 May 1970, flying in the Super Guppy from Mather Air Force Base to Los Alamitos Naval Air Station in a day.

On 2 May 1970 the stage entered storage in Seal Beach but was soon transferred the short distance to the McDonnell Douglas plant at Huntington Beach where it entered temporary storage on 11 December 1970. The stage was finally shipped via Super Guppy from Los Alamitos Naval Air Station to KSC on 25 August 1971, arriving the following day. At KSC it was placed in long term storage on 12 November 1971. A year later, on 28 November 1972, the stage was removed from storage at KSC.

The CSM-117 arrived at KSC on 1 December 1972 and the S-IB-7 stage arrived at KSC on 30 March 1973. The

A S-IVB-200 stage in a Beta test stand at SACTO
(Mid1960s)

S-IB-7 was erected on ML-1 on 28 May 1973, followed by the S-IVB-207 later on the same day. The Instrument Unit for this mission S-IU-208 (which was out of sequence) arrived at KSC on 8 May 1973 and was erected on the launch vehicle on 29 May. Erection of the SLA-23 and CSM-117 followed.

The launch vehicle was transferred to the launch pad 39B on 11 June and the Flight Readiness Test was conducted on 29 June. RP-1 fuel was loaded into the S-IB-7 stage on 11 July and the wet CDDT was performed on 20 July.

The S-IVB-207 formed the second stage of the AS-207 vehicle that launched the second manned crew to the Skylab space station at 0710:50 EDT on 28 July 1973.

The S-IVB-207 stage received the international designation of 1973-50B as it remained in orbit for 0.24 days before re-entering the atmosphere.

S-IVB-208

Summary

Manufactured in 1966/67 and placed in storage until 1972. Exposed to the explosion of the S-IVB-503 stage at SACTO in January 1967. Part of the vehicle that launched the third manned Skylab crew.

S-IVB-208 in final assembly tower at Huntington Beach with S-IVB-210 forward dome in foreground
(9.1966)

S-IVB-208 stage during weight and balance check at Huntington Beach
(29.11.1966)

Engines

The engine configuration for initial build, stage static firing and launch:

Position 201: J-2062

Stage manufacturing

Tank assembly was completed on 18 March 1966. Hydrostatic proof testing of the tank was completed on 22 March 1966. Subsequently the stage was leak and dye-penetrant checked to ensure the structural integrity of its components and welds. The stage was moved into the insulation chamber at Huntington Beach on 5 April 1966 for installation of insulation, which was completed on 25 May.

The J-2062 engine was shipped from Rocketdyne's Canoga Park on 27 April 1966, arriving at Huntington

Static test firing of the S-IVB-208 stage in the Beta I test stand at SACTO
(12.1.1967)

**Static test firing of the S-IVB-208 stage
in the Beta I test stand at SACTO** (12.1.1967)

Beach later the same day. Meanwhile stage joining operations began at Huntington Beach on 15 July 1966. On 10 August 1966 the stage was installed in Tower # 5 at Huntington Beach. Post modification checkout of the J-2 engine was completed on 10 August 1966 and the engine was installed in the S-IVB-208 stage on the same day. Engine alignment was conducted on 24 August. Checkout operations started on 18 August and were completed on 12 October after 39 days of activi-

**The S-IVB-208 (horizontal) and S-IVB-209 stages
in storage in the VCL at SACTO** (1967)

ty. The All-Systems Test was performed on 4 and 5 October.

Stage testing

Post Manufacturing Checkout of the S-IVB-208 stage was completed on 10 November 1966. The stage was weighed on 29 November and found to be 23,749.0 pounds. The stage left Huntington Beach on 2 December 1966, arriving at The Douglas Aircraft Company's SACTO test facility later the same day. The stage had flown in the Super Guppy from Los Alamitos Naval Air Station to Mather Air Force Base. The stage was installed in the Beta I test stand on 5 December and pre-static checkout was completed on 6 January 1967 with a successful simulated acceptance firing being performed on the 5 January which verified the countdown procedure. A final Integrated Systems Test was performed on 12 January 1967 which verified the functional readiness of the stage and facility to proceed with countdown operations prior to static firing.

The acceptance firing countdown was started at 0750 PST on 11 January 1967. The single acceptance firing of 426.6 seconds duration (CD 614076) took place successfully at 1212:16.270 PST on 12 January 1967. Automatic cut-off was initiated because of imminent LOX depletion as detected by the PU processor.

On 20 January 1967 the S-IVB-503 stage located in the nearby Beta III test stand exploded. It was 1,860 feet

**Close up of the S-IVB-208 J-2 engine during the stage
acceptance firing at SACTO's Beta I test stand**
(12.1.1967)

S-IVB-208 stage with technician during the weight and balance measurement at Huntington Beach
(29.11.1966)

The S-IVB-209 (left) and S-IVB-208 (right) stages under long term storage covers in the E&D building at SACTO
(1968)

away from the S-IVB-208 stage at the time. Inspections revealed that the S-IVB-208 stage was not damaged and calculations showed that the pressure wave from the explosion did not cause any problems for the S-IVB-208 stage. It was predicted that the pressure differential at the critical forward dome was 0.16 psid, compared with the buckling pressure of 0.45 psid. Two minor scratches were found in the paint on the forward dome near the LH2 re-pressure connectors, but these scratches were believed not to be associated with the S-IVB-503 incident. The scratches were painted over for corrosion protection. A special leak check was performed to verify the integrity of the stage following the nearby explosion.

The stage was transferred to the Vehicle Checkout Laboratory on 27 January 1967 for post-firing checkout and All-Systems testing. The All-Systems Test was performed on 16 March and verified the basic operation of all the stage systems. Post static checkout was completed on 22 March 1967, following which the stage was placed in long term storage at SACTO. It was eventually transported from SACTO to McDonnell Douglas's Huntington Beach plant on 13 October 1970, flying in the Super Guppy from Mather Air Force Base to Los Alamitos Naval Air Station in a day.

After a year in temporary storage the stage was shipped to KSC, leaving Los Alamitos Naval Air Station on 3 November 1971 and arriving at KSC on 5 November. It was placed in long term storage at KSC on 5 January 1972, before being removed on 28 November 1972.

The S-IB-8 stage was erected on the mobile launcher, ML-1, on 31 July 1973. The S-IVB-208 stage was erected on top of the booster stage on the same day. The out-of-sequence IU, S-IU-207, arrived at KSC on 12 June 1973 and was stacked on the rocket on 1 August 1973. Following this the SLA-24 and CSM-118 were mounted on top of the rocket.

The rocket was transferred to launch pad 39B on 14 August 1973. Flight Readiness Tests were conducted on 5 September and 11 October.

The launch was initially scheduled for 10 November but was delayed as all 8 S-IB-8 tail fins in the S-IB-8 stage had to be replaced. This was after inspection of the tail fins on the rescue booster, S-IB-9 revealed cracks. An inspection of the S-IB-8 stage revealed similar cracks in the fin attachment fittings caused by stress corrosion cracking. The wet CDDT was performed on 2 November. On 7 November the RP-1 was drained in preparation for replacement of the tail fins. The new fins were installed by 13 November and the RP-1 reloaded on the following day.

The S-IVB-208 stage finally was launched as the second stage of the third and final manned Skylab flight, AS-208 Skylab 4, at 0901:23 EST on 16 November 1973.

The S-IVB-208 stage received the international designation of 1973-90B as it remained in orbit for less than half a day before re-entering the atmosphere.

S-IVB-209

Summary

Final S-IVB-200 stage to be static fired. Acted as a potential Skylab and Apollo Soyuz rescue vehicle. Currently on display at KSC.

Engines

The engine configuration for initial build and stage static firing:

Position 201: J-2083

S-IVB-209 enroute to vertical checkout tower # 8 at Huntington Beach for final leak check next to S-IVB-212 cylinder (2.1967)

Stage manufacturing

S-IVB-209 stage tank assembly was completed at Huntington Beach on 8 July 1966. The proof test was completed on 26 July 1966. Insulation and bonding was completed on 6 September 1966. The J-2083 engine was shipped from Rocketdyne's Canoga Park on 20 October 1966, arriving at Huntington Beach later the same day. Post modification checkout of the engine was completed on 10 November 1966 and the engine was installed in the S-IVB-209 stage on 15 November. Factory checkout of the stage was initiated on 28 November 1966 in Tower # 5 in Huntington Beach. This was completed on 26 January 1967, after which installation and replacement of bus modules was accomplished.

Stage testing

Following factory checkout leak checks were completed in Tower # 8 on 10 February. Post Manufacturing Checkout of the S-IVB-209 stage was completed on 16 February 1967 and the stage left Huntington Beach on 9 March 1967, arriving at The Douglas Aircraft Com-

pany's SACTO test facility on the following day. The stage had flown in the Super Guppy from Los Alamitos Naval Air Station to Mather Air Force Base. Pre-static checkout in the Vehicle Checkout Laboratory Tower # 1 took place between 27 March and 13 May. On 14 May the stage was transferred to the Beta I test stand, where it was installed the following day for continuation of pre-firing checkout which concluded on 13 June 1967. The Propulsion System Test took place between 24 and 27 May, the Integrated System Test on 3 June and the simulated static firing took place on 8 and 9 June.

There were two aborted attempts by (the now renamed)

S-IVB-209 stage in the assembly and checkout towers at Huntington Beach
(2.1967)

The S-IVB-209 stage in the insulation chamber at Huntington Beach
(1967)

S-IVB-209 at Huntington Beach
(2.1967)

McDonnell Douglas to test fire the stage on 14 June
(CD 614084). The first abort occurred due to a false
indication from the reusable ignition detect probe. The
second abort was due to an apparent relay failure,
although in post-test investigation it did not repeat the
failure. The stage was subjected to the single acceptance firing of 455.95 seconds duration (CD 614085),
which took place successfully at 1142:05.847 PDT on
20 June 1967, with the countdown having been initiated the day before. The firing was terminated by LOX
depletion cut-off as planned. The Terminal Countdown
had been initiated at 1114 PDT on 20 June. Total propellant consumed in the firing was 189,949 pounds of
LOX and 36,856 pounds of LH2.

Remains of the Beta III test stand at SACTO
(2006)

On 23 June the stage was re-loaded with propellants for
special tests on the LH2 chill-down system and the LH2
depletion sensors. On 26 June 1967 an abbreviated
post-firing checkout of the stage was initiated prior to
removing it from the Beta I test stand. The checkout
was completed on 6 July and the stage was removed
from the Beta I test stand. It was moved to the VCL at
SACTO on 7 July.

As there was no identified mission for this stage in the
foreseeable future, the normal post static checkout was
not performed. Instead the stage was placed in storage
in the Engineering and Development Building at
SACTO on 19 July 1967 until 22 July 1970, when the
stage was transferred by Super Guppy from Mather Air
Force Base to Los Alamitos Naval Air Station for storage in Huntington Beach.

The stage was finally shipped to KSC by Super Guppy,
leaving Los Alamitos Naval Air Station on 11 January
1972 and arriving at KSC the following day. The stage
was placed in long term storage at KSC on 27 March
1972 before being removed on 2 July 1973.

Remains of the Beta I test stand at SACTO
(2006)

Remains of the Beta III test stand at SACTO
(2006)

SA-209 rocket on display at KSC (2006)

S-IVB-210 thrust structure at Huntington Beach
(1966)

On 3 December 1973 the stage was transferred to launch pad 39B atop the S-IB-9 booster stage, and below SLA-22 and CSM-119, forming the SA-209 vehicle which would have acted as a rescue vehicle for the final manned Skylab mission if needed. A Flight Readiness Test was performed on 17 December 1973. However as the Skylab mission was a success SA-209 did not need to be launched and was rolled back into the VAB during February 1974. After disassembly of the stages, the S-IVB-209 stage was placed in long term vertical storage, in low bay cell 1 of the VAB, on 15 April 1974.

In 1975 it was again readied for potential launch, this time as the back-up Apollo Soyuz launch vehicle. However, unlike the Skylab rescue mission, the S-IVB-209 stage was not assembled into a complete Saturn IB stack or transferred to the launch pad. After the success of the Apollo Soyuz mission in 1975 the S-IVB-209 stage was no longer needed as a back up for that mission and has ever since been on display in a horizontal configuration in the KSC Rocket Garden, together with the S-IB-9 booster stage and payload.

S-IVB-210

Summary

The only S-IVB-200 stage to be launched without having had a ground static firing first. Originally built in 1966/67 and then launched as the upper stage of the final manned Apollo mission, the Apollo Soyuz Test Project in 1975. As such, this was the final Saturn stage to be ignited.

Engines

The engine configuration for initial build and launch:

Position 201: J-2087

Stage manufacturing

Stage manufacturing was started on 15 February 1966. The hydrostatic proof test of the tank assembly that would ultimately be used on the S-IVB-210 stage took place on 11 and 12 November 1966. This was followed by a leak check on 16 and 17 November. One minor leak during the proof test, at the LH2 pressure port, was corrected by replacing a cono-seal. The original tank assembly, SN 2010, inadvertently had the common bulkhead filled with filtered city water. The resultant rework to ensure a dry, corrosive-free atmosphere in the common bulkhead caused the tank assembly to fall behind in the manufacturing schedule. Therefore tank assembly SN 2011 (re-designated as SN 2010) was moved forward for allocation to the S-IVB-210 stage.

The J-2087 engine was shipped from Rocketdyne's Canoga Park on 11 November 1966, arriving at Huntington Beach later the same day. Forward and aft skirts

S-IVB-210 aft skirt at Huntington Beach
(1966)

**S-IVB-210 in foreground, S-IVB-505 in background
in vertical checkout tower # 5 at Huntington Beach**
(Early 1967)

and thrust structure installations were completed in Tower # 2 on 17 January 1967 and the stage was moved to Tower # 6 at Huntington Beach on the same day. Post modification checkout of the engine was completed on 12 January 1967 and the engine was installed in the S-IVB-210 stage on 20 January.

Factory stage checkout, initiated on 3 February 1967, was completed on 22 March. However, during this period the stage was removed from the tower for two days (22 and 23 February) to remove and replace helium sphere SN 1157 because it had deformed 0.020 inch. It was replaced by sphere SN 1134. At the same time it was decided to replace helium sphere SN 1146 and replace it with sphere SN 1154. This was because of doubts about the integrity of SN 1146. Engine alignment was satisfactorily accomplished on 7 March 1967. The All Systems Test was initiated on 13 March but was not completed until 14 March because four attempts were needed because of problems encountered. The stage was removed from the VCL on 27 March 1967 after 34 working days in the tower.

Subsequent to VCL checkout at Huntington Beach the stage was placed in Tower # 8 for the propellant tanks system leak check. This was performed on 3 April, to ensure the leak-free condition of the tank assembly.

Stage testing

Due to the lack of an identified mission no further testing was performed on the stage at this time and it was placed in long term storage in Building 22 at Huntington Beach from 23 April 1967 to mid-1970. Initially, post-checkout modifications were undertaken, which were completed on 14 September 1967. After an inspection in mid-1970 it was returned to storage in Huntington Beach on 21 August 1970, before being removed on 27 January 1971. Post Manufacturing Checkout of the S-IVB-210 stage finally was completed on 11 January 1972 and the stage left Huntington Beach on 7 November 1972, arriving at KSC on the following day. The stage had flown in the Super Guppy from Los Alamitos Naval Air Station to KSC. It had been decided, as a cost-cutting measure that this, and subsequent S-IVB-200 stages would not be static fired at SACTO.

The stage was placed in long term storage at KSC from 14 December 1972 until 11 February 1974. It was returned to storage on 29 May 1974 and was removed on 3 September 1974.

On 13 January 1975 the S-IB-10 stage was erected on the mobile launcher. The second stage was stacked the following day, and the IU and boilerplate spacecraft on 16 January. The CSM-111 was mated to the SLA-18 on 5 March. Between 11 and 19 March cracked fins on the S-IB-10 stage were replaced. After removal of the boilerplate spacecraft the flight spacecraft was erected on the launch vehicle on 19 March 1975.

On 24 March 1975 the AS-210 vehicle was rolled to

S-IVB-210 after final painting at Huntington Beach
(4.1967)

**S-IVB-210 in background, S-IVB-507 components
in foreground at Huntington Beach**
(4.1967)

launch pad 39B. On top of the booster stage was the S-IVB-210 stage, the S-IU-210, SLA-18, DM-2 and the CSM-111. The CDDT took place on 3 July 1975. The vehicle was launched at 1550:00 EDT on 15 July 1975 with S-IVB-210 forming the second stage of the AS-210 launch vehicle that orbited the final manned Apollo capsule used in the Apollo Soyuz Test Project. The S-IVB-210 stage was only the second Saturn stage (after the Apollo 17 third stage) to be launched without having had an acceptance static firing on the ground.

The S-IVB-210 stage received the international designation of 1975-66B as it remained in orbit for one day before re-entering the atmosphere.

S-IVB-211

Summary

The original construction included internal fire-retardant material as it was under consideration for the manned Skylab workshop. Stage was never static fired and is currently on display, without its engine, at the USSRC. The stage has recently been investigated in support of the Ares design activity. The engine is in a museum in Germany.

Engines

The engine configuration for initial build:

Position 201: J-2095

Protective storage cover being placed over the S-IVB-211 stage in the VCL at SACTO
(1967)

Stage manufacturing

The S-IVB-211 tank insulation and clip bonding was completed on 7 February 1967 in the Insulation Chamber at Huntington Beach. The stage was moved to Tower # 4 for cleaning and then returned to the Insulation Chamber for installation of the liquid hydrogen tank fire retardant liner which was completed on 12 March. This was installed in preparation for the planned, but not realized, use of this stage as an orbital workshop. The stage was then moved to building 42 for cleaning.

Technician removes vacuum bag from the interior of the S-IVB-211 stage LH2 tank after bonding of the aluminum liner (3.1967)

S-IVB-211 forward skirt undergoing cold plate checkout at Huntington Beach. S-IVB-506 aft skirt in background
(1966)

**S-IVB-211 being unloaded from the Super
Guppy at Mather Air Force Base**
(18.10.1968) 6870115

The S-IVB-211 stage on display at USSRC
(2004)

The J-2095 engine was shipped from Rocketdyne's Canoga Park on 3 January 1967, arriving at Huntington Beach later the same day. Post modification checkout of the engine was completed on 1 March 1967 and the engine was installed in the S-IVB-211 stage on 11 April. Joining of the thrust structure and forward and aft skirts was initiated on 23 March. The S-IVB-211 stage entered post-manufacturing checkout on 19 April 1967 following stage joining. On 6 July 1967 painting of the S-IVB-211 stage was completed at Huntington Beach and the stage was ready to depart for SACTO. However, it was decided to place the stage in long-term storage at Huntington Beach from 3 August 1967.

Stage testing

Post Manufacturing Checkout of the S-IVB-211 stage was completed until 11 October 1968 and the stage left Huntington Beach on 17 October 1968, arriving at the SACTO test facility on the following day. The stage had flown in the Super Guppy from Los Alamitos Naval Air Station to Mather Air Force Base.

Due to the lack of an identified mission no further testing was performed on the stage at this time and it was placed in long term storage at SACTO on 18 December 1968 for the following 2 years, without any plans for a static firing. The stage was flown back to Huntington Beach on 15 September 1970 prior to completion of modifications, checkouts or static firings at SACTO. Travel was via the Super Guppy from Mather Air Force Base to Los Alamitos Naval Air Station. On 5 October 1970 the stage was transported the short distance across to NAA at Seal Beach for long term storage, where it remained until 2 September 1971 when it was returned to Huntington Beach. McDonnell Douglas performed an expanded All Systems Test on the stage which was completed on 28 April 1972.

The S-IVB-211 stage was flown to KSC on 26 June 1972 on the Super Guppy aircraft, leaving Los Alamitos Naval Air Station and arriving at KSC the following day, 27 June. The stage was placed in long term storage at KSC on 29 June 1972. At some time subsequently the stage was sent to Huntsville. For a time in the late 1970s the S-IVB-211 stage was displayed horizontally at MSFC along with the S-IB-11 stage.

At a later date the S-IVB-211 stage was transported to the Alabama Space and Rocket Centre (now USSRC) in Huntsville, Alabama where it was added to an outdoor display together with other spare hardware and mock-ups to simulate the Skylab space station. In more recent years the J-2 engine (J-2095) attached to this stage was removed and sent for display to the Herman Oberth museum in Germany. The stage itself remained virtually untouched until 2007 when NASA engineers removed a number of components from the stage and tested them at MSFC in the frame of the Ares design activity.

The S-IVB-211 stage on display at USSRC
(2008)

S-IVB-212 (Skylab Orbital Workshop)

Summary

Final S-IVB-200 stage, converted for use and flown as Skylab OWS. Originally built with a J-2 engine but never test fired at stage level. Engine removed as part of the conversion. Re-entered the atmosphere in 1979. Elements of the engine still being used by NASA for Ares testing in 2008.

Engines

The initial engine configuration was:

Position 301: J-2103

Stage manufacturing

The twelfth second stage for the Saturn IB family of rockets was designated S-IVB-212. It was assembled by McDonnell Douglas at its plant on the Los Angeles coast at Huntington Beach. Subassemblies for this stage were manufactured at McDonnell Douglas facilities in Santa Monica and Tulsa, starting in late 1966. By early 1967 the propellant tank assembly had been completed comprising the forward and lower domes, the cylindrical fuel tank and the common bulkhead.

The interior walls of the hydrogen tank were covered in tiles to insulate the cold liquid hydrogen. The tank assembly was then ready for the two main manufacturing tests, proof pressure and leakage, which were performed in order to verify the structural integrity of the stage propellant tank assembly.

S-IVB-212 in vertical assembly weld position at Huntington Beach, S-IVB-210 in foreground
(Early 1967)

Hydrostatic proof pressure testing was performed on 8 and 9 March 1967, followed by leakage checks on 14 and 17 March. The proof pressure test saw the LOX tank tested to 51.0 psi and the fuel tank to 38.0 psi. Pressures were applied for 5 minutes at each level. The leakage tests were performed with gaseous nitrogen at 12.0 psi. After 10 minutes the pressure had dropped to 11.9 psi which was deemed an acceptable leak rate. Finally, Freon gas was applied to the interior of the tanks and detectors were placed at joints and welds to see if any gas had escaped. The stage passed these tests and continued to final assembly. The aft skirt, forward skirt and thrust structure assemblies were attached to the tank assembly at the end of March 1967.

S-IVB-212 LOX tank assembly
(Early 1967)

S-IVB-212 by the assembly tower at Huntington Beach
(9.1967)

**The J-2 engine is mated to the
S-IVB-212 stage at Huntington Beach**
(Early 1967) 6757790

**J-2X engine, with powerpack from the S-IVB-212 J-2
engine, being tested in support of the Ares project at SSC**
(15.2.2008)

Meanwhile, in parallel, the J-2 engine allocated to this stage, J-2103, was being tested and delivered. The engine was manufactured and assembled by Rocketdyne at Canoga Park starting in the second quarter of 1966. It was subjected to 3 acceptance firings totaling 615.4 seconds at the Santa Susana Field Laboratory (SSFL), with the final firing being on 25 January 1967. The engine was accepted by NASA on 22 February 1967 and shipped by truck across Los Angeles to Huntington Beach on the following day. The engine was installed in the S-IVB-212 stage following the assembly of the thrust structure.

The stage was fully equipped with various equipment and was transferred to Vehicle Checkout Laboratory (VCL) tower number 6 at the Huntington Beach plant on 29 June 1967.

Stage testing

On 25 July the stage was declared complete and ready to enter systems checkout. These tests were performed between 25 July and 14 September 1967 consuming 37 working days. Power was first applied to the stage on 1 August 1967. Some tests were not accepted as closed until 22 September. The successful completion of these tests would clear the stage for shipment to the Sacramento Test Facility (SACTO) for static firing.

On 14 September 1967 McDonnell Douglas concluded the All Systems Test that ended systems checkout of the S-IVB-212 stage. This test demonstrated the combined operation of the stage electrical, hydraulic, propulsion, instrumentation and telemetry systems under simulated flight conditions. A total of 33 checkout procedures involving the stage systems were accomplished during this period. The final procedure secured the forward skirt thermo-conditioning system following VCL automatic checkout activities and consisted of a system cleanliness check, a drain and dry procedure, a leak check and preparations for stage shipment.

The stage was removed from the VCL on 18 September 1967 and was moved to tower # 8 for a production acceptance leak test of the propellant tanks assembly using helium gas. The LOX and fuel tanks were pressurized to 10.0 psi. This was accomplished between 25 and 27 September 1967. Six leaks exceeded the limit of 0.001 cc/s but were rectified by replacing seals.

On 18 October 1967 McDonnell Douglas completed painting and final inspection of the S-IVB-212 stage and prepared it for storage at the Space Systems Center at Huntington Beach. As there was no mission identified for this stage it was not sent to SACTO for static firing. The stage remained in storage at Huntington Beach from 3 November 1967 until March 1969. At that time a decision was made to refurbish the S-IVB-212 stage and use it as the Skylab OWS. This was a major reconfiguration that was to take three years to complete.

On 26 March 1969 McDonnell Douglas shipped the inter-stage for S-IVB-212 to MAF. In April 1969 S-IVB-212 was moved from the vertical checkout tower # 8 to tower # 6, where the J-2 engine was disconnected as part of the refurbishment of the stage. The engine was transferred to engineering at Huntington Beach in support of S-IVB-210 and S-IVB-211. Engine checkout was completed on 2 September 1971 at Huntington Beach. The engine was shipped to Edwards Field Laboratory for storage on 27 January 1972. It was then shipped to KSC on 25 March 1974 for storage. It was placed in storage with the S-IVB-513 stage on 5 June 1974. It was taken out of storage on 23 October 1974 and checkout was completed on 2 December 1974. Finally the engine was stripped down for spares. The power pack from the engine was later used by Rocketdyne to support their linear aerospike engine testing and more recently in 2007 and 2008 the power pack was used in early Ares J-2X testing at Stennis Space Center.

In May 1969 the forward skirt was removed from the

stage. The stage, now consisting only of a tank assembly, was placed in the insulation chamber for removal of the insulation tiles. In June 1969 the tank assembly was removed from the insulation chamber and transported to the vertical assembly and checkout building.

During November 1969 S-IVB components were removed from the thrust structure and aft skirt. During the course of the next two and a half years the tank interior was configured for manned habitation including the installation of the living accommodation floors and walls. The forward and aft skirts were re-attached to the stage in May 1971. In place of the J-2 engine, at the aft end of the OWS, a Thruster Attitude Control System (TACS) module and radiator were installed in June 1971. McDonnell Douglas completed the conversion of the S-IVB-212 stage to the Skylab Orbital Workshop in the summer of 1972.

The Skylab Orbital Workshop, together with the payload shroud, was transferred from the McDonnell Douglas plant in Huntington Beach to the nearby Seal Beach Naval docks where they were loaded on board the AKD Point Barrow. This ship departed on 8 September 1972 for the journey through the Panama Canal and round the tip of Florida before arriving at the Kennedy Space Center on 22 September 1972.

The S-IVB-212 stage was eventually launched in the modified Skylab configuration on 14 May 1973. It was host to three visiting astronaut crews before re-entering on 11 July 1979 after 2,248 days in orbit. It had been given the orbital designation of 1973-27A.

S-IVB SACTO Battleship

Config/Test	Countdown number	Test stand	Date	Planned duration	Actual duration	Engine	Comments
IB/non-firing	CD 614000	SACTO/Beta I	18.9.64.	N/A	N/A	J-2003	LN2 and LH2 cryogenic loading
IB/non-firing	CD 614002	SACTO/Beta I	25.9.64.	N/A	N/A	J-2003	LOX and LH2 propellant loading
IB/non-firing	CD 614003	SACTO/Beta I	2.10.64.	N/A	N/A	J-2003	Engine chilldown test
IB/non-firing	CD 614003	SACTO/Beta I	9.10.64.	N/A	N/A	J-2003	Thrust chamber jacket chill down test
IB/firing attempt	CD 614004	SACTO/Beta I	24.10.64.	10s	0s	J-2003	Start tank blowdown test, failed ignition
IB/firing attempt	CD 614005	SACTO/Beta I	7.11.64.	10s	0s	J-2003	3 aborted attempts at a 10s firing
IB/non-firing	CD 614006	SACTO/Beta I	24.11.64.	N/A	N/A	J-2003	Gas generator ignition test
IB/firing	CD 614007	SACTO/Beta I	1.12.64.	10s	10.67s	J-2003	Successful shakedown firing
IB/firing	CD 614008	SACTO/Beta I	9.12.64.	50s	50.7s	J-2003	Successful shakedown firing
IB/firing	CD 614009	SACTO/Beta I	15.12.64.	150s	150.4s	J-2003	Successful shakedown firing
IB/firing	CD 614010	SACTO/Beta I	23.12.64.	Full duration	414.6s	J-2003	Successful nearly full duration firing
IB/non-firing	CD 614011	SACTO/Beta I	8.1.65.	N/A	N/A	J-2003	J-2 engine temp conditioning test
IB/non-firing	CD 614012	SACTO/Beta I	14.1.65.	N/A	N/A	J-2003	J-2 engine temp conditioning test
IB/non-firing	CD 614013	SACTO/Beta I	16.1.65.	N/A	N/A	J-2003	J-2 engine temp conditioning test
IB/non-firing	CD 614014	SACTO/Beta I	9.2.65.	N/A	N/A	J-2013	J-2 engine temp conditioning test
IB/non-firing	CD 614015	SACTO/Beta I	17.2.65.	N/A	N/A	J-2013	J-2 engine temp conditioning test
IB/non-firing	CD 614016	SACTO/Beta I	18.2.65.	N/A	N/A	J-2013	J-2 engine temp conditioning test
IB/non-firing	CD 614017	SACTO/Beta I	25.2.65.	N/A	N/A	J-2013	J-2 engine temp conditioning test
IB/non-firing	CD 614018	SACTO/Beta I	2.3.65.	N/A	N/A	J-2013	J-2 engine temp conditioning test
IB/non-firing	CD 614019	SACTO/Beta I	6.3.65.	N/A	N/A	J-2013	J-2 engine temp conditioning test
IB/firing	CD 614020	SACTO/Beta I	13.3.65.	10s	11.8s	J-2013	Successful shakedown firing
IB/firing	CD 614021	SACTO/Beta I	19.3.65.	Full duration	29.2s	J-2013	Full duration attempt aborted due to instrumentation problem
IB/firing attempt	CD 614022	SACTO/Beta I	25.3.65.	Full duration	0s	J-2013	High EMR, PU excursion firing – 3 aborted attempts
IB/firing	CD 614023	SACTO/Beta I	31.3.65.	Full duration	470s	J-2013	Successful full duration high EMR, PU excursion firing
IB/firing	CD 614024	SACTO/Beta I	7.4.65.	Full duration	42s	J-2013	Low EMR, PU excursion firing. Aborted due to instrumentation problem
IB/firing	CD 614025	SACTO/Beta I	15.4.65.	Full duration	506.75s	J-2013	Successful low EMR, PU excursion firing
IB/non-firing	CD 614026	SACTO/Beta I	22.4.65.	N/A	N/A	J-2013	Spring rate simulator verification and ambient gimbal test
IB/firing	CD 614028	SACTO/Beta I	27.4.65.	Full duration	374s	J-2013	High EMR firing terminated due to high temperature
IB/firing	CD 614030	SACTO/Beta I	4.5.65.	Full duration	493.5s	J-2013	Hot gimbal, full duration. Successful
IB/non-firing	CD 614031	SACTO/Beta I	13.5.65.	N/A	N/A	J-2013	Aft interstage environmental test
IB/non-firing	CD 614032	SACTO/Beta I	14.5.65.	N/A	N/A	J-2013	Aft interstage environmental test
V/firing	CD 614033	SACTO/Beta I	19.6.65.	First burn, coast, restart	8.92s	J-2020	Aborted due to facility logic problem
V/firing	CD 614034	SACTO/Beta I	26.6.65.	First burn, coast, second burn	167+3.84s	J-2020	94 min coast time between burns. 2nd burn aborted due to instrumentation noise
V/firing	CD 614035	SACTO/Beta I	1.7.65.	First burn, coast, second burn	5.45+1.72s	J-2020	Ist firing aborted due to control logic. 2nd firing aborted due to fire and explosion
V/firing attempt	CD 614041	SACTO/Beta I	12.8.65.	First burn, coast, second burn	0s	J-2020	Aborted due to leakage in loading
V/firing	CD 614042	SACTO/Beta I	13.8.65.	First burn, coast, second burn	16s	J-2020	Terminated early due to minor fire
V/firing	CD 614043	SACTO/Beta I	17.8.65.	First burn, coast, second burn	170+319s	J-2020	92 min coast time between burns. 2nd burn gimbal
V/firing	CD 614044	SACTO/Beta I	20.8.65.	First burn, coast, second burn	170.9+360.2s	J-2020	41 min coast time between firings. 2nd burn gimbal

SA-T Tests

config.	no.		Time	Actual	101	102	103	104	105	106	107	108	Comments
SA-T	SAT-01	MSFC/STTE	28.3.60. 1058	8s						H-1007		H-1009	2 engine test - successful
SA-T	SAT-02	MSFC/STTE	6.4.60. 1115	7s					H-1006	H-1007	H-1008	H-1009	4 engine test - successful
SA-T	SAT-03	MSFC/STTE	29.4.60. 1728	8s	H-1001	H-1002	H-1003	H-1005	H-1006	H-1007	H-1008	H-1009	First 8 engine test - successful
SA-T	SAT-04	MSFC/STTE	17.5.60. 1732	25s	H-1001	H-1002	H-1003	H-1005	H-1006	H-1007	H-1008	H-1009	Successful
SA-T	SAT-05	MSFC/STTE	26.5.60. 1703	35s	H-1001	H-1002	H-1003	H-1005	H-1006	H-1007	H-1008	H-1009	Successful
SA-T	SAT-06	MSFC/STTE	3.6.60. 1656	75s	H-1001	H-1002	H-1003	H-1005	H-1006	H-1007	H-1008	H-1009	Successful
SA-T	SAT-07	MSFC/STTE	8.6.60. 1656	110s	H-1001	H-1002	H-1003	H-1005	H-1006	H-1007	H-1008	H-1009	Successful
SA-T	SAT-08	MSFC/STTE	15.6.60. 1709	121s	H-1001	H-1002	H-1003	H-1005	H-1006	H-1007	H-1008	H-1009	Successful
SA-T1	SAT-09	MSFC/STTE	2.12.60. 1801	1.7s	H-1001	H-1002	H-1003	H-1005	H-1006	H-1007	H-1008	H-1009	Low temperature abort
SA-T1	SAT-10	MSFC/STTE	10.12.60. 1449	6s						H-1007		H-1009	2 engine test - successful
SA-T1	SAT-11	MSFC/STTE	20.12.60. 1639	61s	H-1001	H-1002	H-1003	H-1005	H-1006	H-1007	H-1008	H-1009	Successful
SA-T1	SAT-12	MSFC/STTE	31.1.61. 1648	113s	H-1001	H-1002	H-1003	H-1005	H-1006	H-1007	H-1008	H-1009	Successful
SA-T1	SAT-13	MSFC/STTE	14.2.61. 1648	109s	H-1001	H-1002	H-1003	H-1005	H-1006	H-1007	H-1008	H-1009	Successful
SA-T2	SAT-14	MSFC/STTE	27.6.61. 1551	29.9s	H-1001	H-1002	H-1003	H-1005	H-1006	H-1007	H-1008	H-1009	New LOX domes
SA-T2	SAT-15	MSFC/STTE	7.7.61. 1620	118s	H-1001	H-1002	H-1003	H-1005	H-1006	H-1007	H-1008	H-1009	Successful
SA-T2	SAT-16	MSFC/STTE	18.7.61. 1441	116s	H-1001	H-1002	H-1003	H-1005	H-1006	H-1007	H-1008	H-1009	In-flight cut-off sequence tested
SA-T2	SAT-17	MSFC/STTE	3.8.61. 1709	1.2s	H-1001	H-1002	H-1003	H-1005	H-1006	H-1007	H-1008	H-1009	Successful
SA-T2	SAT-18	MSFC/STTE	7.8.61. 1543	123s	H-1001	H-1002	H-1003	H-1005	H-1006	H-1007	H-1008	H-1009	Successful
SA-T2	SAT-19	MSFC/STTE	25.8.61. 1542	114s	H-1001	H-1002	H-1003	H-1005	H-1006	H-1007	H-1008	H-1009	Successful
SA-T3	SAT-20	MSFC/STTE	30.11.61. 1701	95s	H-1001	H-1002	H-1003	H-1005	H-1006	H-1007	H-1008	H-1009	Fire indication abort
SA-T3	SAT-21	MSFC/STTE	19.12.61. 1542	68s	H-1001	H-1002	H-1003	H-1005	H-1006	H-1007	H-1008	H-1009	Successful
SA-T3	SAT-22	MSFC/STTE	18.1.62. 1639	122s	H-1001	H-1002	H-1003	H-1005	H-1006	H-1007	H-1008	H-1009	Successful
SA-T3	SAT-23	MSFC/STTE	6.2.62. 1641	46s	H-1001	H-1002	H-1003	H-1005	H-1006	H-1007	H-1008	H-1009	Successful
SA-T3	SAT-24	MSFC/STTE	20.2.62. 1640	55s	H-1001	H-1002	H-1003	H-1005	H-1006	H-1007	H-1008	H-1009	Fire indication abort
SA-T4	SAT-25	MSFC/STTE	19.6.62. 1646	32s	H-1001	H-1002	H-1003	H-1005	H-1006	H-1007	H-1008	H-1009	Successful
SA-T4	SAT-26	MSFC/STTE	12.7.62. 1640	12.04s	H-1001	H-1002	H-1003	H-1005	H-1006	H-1007	H-1008	H-1009	Instrumentation abort
SA-T4	SAT-27	MSFC/STTE	13.7.62. 1656	20.43s	H-1001	H-1002	H-1003	H-1005	H-1006	H-1007	H-1008	H-1009	Instrumentation abort
SA-T4	SAT-28	MSFC/STTE	17.7.62. 1640	120.00s	H-1001	H-1002	H-1003	H-1005	H-1006	H-1007	H-1008	H-1009	Successful
SA-T4.5	SAT-29	MSFC/STTE	26.10.62. 1651	31.00s	H-5001	H-5006	H-5011	H-5014	H-2009	H-2011	H-2013	H-2014	Block II engines
SA-T4.5	SAT-30	MSFC/STTE	2.11.62. 1644	65.00s	H-5001	H-5006	H-5011	H-5014	H-2009	H-2011	H-2013	H-2014	Successful
SA-T4.5	SAT-31	MSFC/STTE	9.11.62. 1642	115.00s	H-5001	H-5006	H-5011	H-5014	H-2009	H-2011	H-2013	H-2014	Successful

S-1 Tests

Stage	Firing no.	Test stand	Date	Local Time	Duration Planned	Duration Actual	101	102	103	104	105	106	107	108	Comments
S-I-1	SA-01	MSFC/STTE	29.4.61.	1638	Calibration	30s	H-1016	H-1017	H-1019	H-1021	H-1011	H-1012	H-1013	H-1015	Successful
S-I-1	SA-02	MSFC/STTE	5.5.61.	1612	Full duration	44.17s	H-1016	H-1017	H-1019	H-1021	H-1011	H-1012	H-1013	H-1015	Terminated early due to leak
S-I-1	SA-03	MSFC/STTE	11.5.61.	1548	Full duration	111s	H-1016	H-1017	H-1019	H-1021	H-1011	H-1012	H-1013	H-1015	Successful
S-I-2	SA-04	MSFC/STTE	10.10.61.	1659	Calibration	32s	H-1032	H-1033	H-1034	H-1038	H-1028	H-1029	H-1035	H-1036	Successful
S-I-2	SA-05	MSFC/STTE	24.10.61.	1641	Full duration	119s	H-1032	H-1033	H-1034	H-1038	H-1028	H-1029	H-1035	H-1036	Successful
S-I-3	SA-06	MSFC/STTE	10.4.62.	1713	Calibration	31s	H-1045	H-1047	H-1048	H-1049	H-1040	H-1041	H-1042	H-1043	Defective turbo pump bearing
S-I-3	SA-07	MSFC/STTE	17.5.62.	1642	Calibration	31.02s	H-1045	H-1047	H-1048	H-1049	H-1040	H-1041	H-1042	H-1043	Successful
S-I-3	SA-08	MSFC/STTE	24.5.62.	1642	Full duration	119.43s	H-1045	H-1047	H-1048	H-1049	H-1040	H-1041	H-1042	H-1043	Successful
S-I-4	SA-09	MSFC/STTE	11.9.62.	1117	Calibration	31.53s	H-1054	H-1055	H-1056	H-1057	H-1051	H-1052	H-1053	H-1058	Successful
S-I-4	SA-10	MSFC/STTE	26.9.62.	1642	Full duration	121.5s ib/127.43s ob	H-1054	H-1055	H-1056	H-1057	H-1051	H-1052	H-1053	H-1058	Successful
S-I-5	SA-11	MSFC/STTE	27.2.63.	1647	Calibration	31.96s	H-5002	H-5003	H-5004	H-5005	H-2001	H-2002	H-2003	H-2004	Successful
S-I-5	SA-12	MSFC/STTE	13.3.63.	1617	Full duration	144.44s	H-5002	H-5003	H-5004	H-5005	H-2001	H-2002	H-2003	H-2004	Propulsion system anomolies
S-I-5	SA-13	MSFC/STTE	27.3.63.	1640	Full duration	143.47s	H-5002	H-5003	H-5004	H-5005	H-2001	H-2002	H-2003	H-2004	Successful
S-I-6	SA-14	MSFC/STTE	15.5.63.	1640	Calibration	33.75s	H-5007	H-5008	H-5009	H-5010	H-2005	H-2006	H-2008	H-2009	Fire after valve failed to close
S-I-6	SA-15	MSFC/STTE	6.6.63.	1642	Full duration	142.37s	H-5007	H-5008	H-5009	H-5010	H-2005	H-2006	H-2008	H-2007	Successful
S-I-7	SA-16	MSFC/STTE	2.10.63.	1638	Calibration	33.78s	H-5013	H-5014	H-5015	H-5016	H-2010	H-2011	H-2012	H-2015	Successful
S-I-7	SA-17	MSFC/STTE	22.10.63.	1639	Full duration	138.93s ib/145s ob	H-5013	H-5027	H-5015	H-5016	H-2010	H-2011	H-2012	H-2015	Successful
S-I-8	SA-20	MSFC/STTE	26.5.64.	1642:23	Calibration	48.94s ib	H-5019	H-5020	H-5021	H-5022	H-2016	H-2017	H-2018	H-2029	Successful
S-I-8	SA-21	MSFC/STTE	11.6.64.	1639:55	Full duration	139.92s ib/145.61s ob	H-5019	H-5020	H-5021	H-5022	H-2016	H-2017	H-2018	H-2029	Successful
S-I-9	SA-18	MSFC/STTE	13.3.64.	1634:50	Calibration	35.22s ib	H-5023	H-5012	H-5025	H-5026	H-2020	H-2022	H-2023	H-2024	Successful
S-I-9	SA-19	MSFC/STTE	24.3.64.	1335:33	Full duration	142.21s ib	H-5023	H-5012	H-5025	H-5026	H-2020	H-2022	H-2023	H-2024	Successful
S-I-10	SA-22	MSFC/STTE	22.9.64.	1637:57	Calibration	3.01s	H-5028	H-5029	H-5030	H-5031	H-2034	H-2026	H-2027	H-2030	Lack of thrust signal
S-I-10	SA-23	MSFC/STTE	24.9.64.	1637:07	Calibration	35.08s	H-5028	H-5029	H-5030	H-5031	H-2034	H-2026	H-2027	H-2030	Successful
S-I-10	SA-24	MSFC/STTE	6.10.64.	1638:34	Full duration	149.93s ib/154.48s ob	H-5028	H-5029	H-5030	H-5031	H-2034	H-2026	H-2027	H-2030	Successful

S-IV Battleship

Stage	Firing no.	Test stand	Date	Local Time	Duration Actual	Engine type	Comments
S-IV BS	1	SACTO/Alpha TS1	17.8.62.	1112	10s	RL10A-1	Successful
S-IV BS	2	SACTO/Alpha TS1	7.9.62.	1550	13.6s	RL10A-1	Aborted due to LOX valve failure
S-IV BS	3	SACTO/Alpha TS1	15.9.62.	1506	28.3s	RL10A-1	Aborted due to high water temp
S-IV BS	4	SACTO/Alpha TS1	24.9.62.		60s	RL10A-1	Successful
S-IV BS	5	SACTO/Alpha TS1	29.9.62.		42s	RL10A-1	Aborted due to low ullage pressure
S-IV BS	6	SACTO/Alpha TS1	1.10.62.		7.2s	RL10A-1	Aborted due to pressure leak
S-IV BS	7	SACTO/Alpha TS1	4.10.62.		420s	RL10A-1	Successful
S-IV BS	8	SACTO/Alpha TS1	30.10.62.	1702	70s	RL10A-1	Aborted due to instrumentation problem
S-IV BS	9	SACTO/Alpha TS1	3.11.62.		448s	RL10A-1	Successful
S-IV BS	10	SACTO/Alpha TS1	8.11.62.		38.5s	RL10A-1	Aborted due to fire
S-IV BS	1	SACTO/Alpha TS1	26.1.63.		468s	RL10A-3	Successful
S-IV BS	2	SACTO/Alpha TS1	25.2.63.		6.5s	RL10A-3	Aborted due to fire
S-IV BS	3-16	SACTO/Alpha TS1				RL10A-3	Various firings
S-IV BS	17	SACTO/Alpha TS1	4.5.63.		444s	RL10A-3	Successful

S-IB Tests

Stage	Firing no.	Test stand	Date	Local Time	Duration Planned	Duration Actual	Engines 101	102	103	104	105	106	107	108	Comments
S-IB-1	SA-25	MSFC/STTE	1.4.65.	1702:15	Calibration	35.174s ib/35.294s ob	H-7046	H-7047	H-7048	H-7049	H-4044	H-4045	H-4046	H-4047	Successful
S-IB-1	SA-26	MSFC/STTE	13.4.65.	1638:13	Full duration	138.210s ib/145.010s ob	H-7046	H-7047	H-7048	H-7049	H-4044	H-4045	H-4046	H-4047	Thrust chamber tube leak
S-IB-2	SA-27	MSFC/STTE	8.7.65.	1641:25	Calibration	3.002s ib/3.123s ob	H-7051	H-7052	H-7050	H-7054	H-4048	H-4049	H-4050	H-4051	Abort caused by thrust OK switch
S-IB-2	SA-28	MSFC/STTE	9.7.65.	1636:24	Calibration	35.192s ib/35.302s ob	H-7051	H-7052	H-7050	H-7054	H-4048	H-4049	H-4050	H-4051	Successful
S-IB-2	SA-29	MSFC/STTE	20.7.65.	1435:59	Full duration	143.285s ib/144.282s ob	H-7051	H-7052	H-7050	H-7054	H-4048	H-4049	H-4050	H-4051	Post-test tank ripples
S-IB-3	SA-30	MSFC/STTE	12.10.65.	1643	Calibration	35.295s	H-7056	H-7058	H-7059	H-7060	H-4053	H-4054	H-4056	H-4057	Successful
S-IB-3	SA-31	MSFC/STTE	26.10.65.	1641	Full duration	146.226s	H-7056	H-7058	H-7059	H-7060	H-4053	H-4054	H-4056	H-4057	Successful
S-IB-4	SA-32	MSFC/STTE	17.1.66.	1644:11	Calibration	35.227s ib/35.339s ob	H-7062	H-7063	H-7064	H-7065	H-4058	H-4059	H-4060	H-4061	Successful
S-IB-4	SA-33	MSFC/STTE	21.1.66.	1640:20	Full duration	143.934s ib/147.110s ob	H-7062	H-7063	H-7064	H-7065	H-4058	H-4059	H-4060	H-4061	Successful
S-IB-5	SA-34	MSFC/STTE	23.3.66.	1749	Calibration	35s	H-7066	H-7067	H-7068	H-7069	H-4063	H-4064	H-4065	H-4066	Successful
S-IB-5	SA-35	MSFC/STTE	31.3.66.	1641	Full duration	144.6s	H-7066	H-7067	H-7068	H-7069	H-4063	H-4064	H-4065	H-4066	Successful
S-IB-6	SA-36	MSFC/STTE	23.6.66.	1639:57	Calibration	35.464s ib/35.580s ob	H-7071	H-7072	H-7073	H-7075	H-4068	H-4069	H-4070	H-4071	Successful
S-IB-6	SA-37	MSFC/STTE	29.6.66.	1640:00	Full duration	138.580s ib/141.236s ob	H-7071	H-7072	H-7073	H-7075	H-4068	H-4069	H-4070	H-4071	Damaged thrust chamber tubes
S-IB-7	SA-38	MSFC/STTE	1.9.66.	1659	Calibration	35s	H-7077	H-7078	H-7076	H-7080	H-4073	H-4074	H-4075	H-4076	Successful
S-IB-7	SA-39	MSFC/STTE	13.9.66.	1650	Full duration	140s	H-7077	H-7078	H-7076	H-7080	H-4073	H-4074	H-4075	H-4076	Successful
S-IB-8	SA-40	MSFC/STTE	16.11.66.	1640:04	Calibration	35.444s ib/35.560s ob	H-7082	H-7083	H-7081	H-7085	H-4077	H-4078	H-4079	H-4080	Turbine wheel blade damage
S-IB-8	SA-41	MSFC/STTE	29.11.66.	1640:54	Full duration	142.656s ib/145.352s ob	H-7082	H-7083	H-7081	H-7085	H-4077	H-4071	H-4079	H-4080	Successful
S-IB-9	SA-42	MSFC/STTE	24.2.67.	1708:06	Calibration	13.252s ib/13.528s ob	H-7090	H-7087	H-7088	H-7089	H-4082	H-4086	H-4084	H-4085	Instrumentation abort
S-IB-9	SA-43	MSFC/STTE	27.2.67.	1641:44	Calibration	35.324s ib/35.440s ob	H-7090	H-7087	H-7088	H-7089	H-4082	H-4086	H-4084	H-4085	Successful
S-IB-9	SA-44	MSFC/STTE	7.3.67.	1545:43	Full duration	3.080s ib/3.356s ob	H-7090	H-7087	H-7088	H-7089	H-4082	H-4086	H-4084	H-4085	Instrumentation abort
S-IB-9	SA-45	MSFC/STTE	7.3.67.	1821:57	Full duration	142.400s ib/145.448s ob	H-7090	H-7087	H-7088	H-7089	H-4082	H-4086	H-4084	H-4085	Successful despite engine damage
S-IB-10	SA-46	MSFC/STTE	9.5.67.	1705:31	Calibration	35.308s ib/35.424s ob	H-7091	H-7099	H-7093	H-7094	H-4087	H-4088	H-4089	H-4090	Successful
S-IB-10	SA-47	MSFC/STTE	22.5.67.	1733:55	Full duration	142.712s ib/145.712s ob	H-7091	H-7099	H-7093	H-7094	H-4087	H-4088	H-4089	H-4090	Successful
S-IB-11	SA-48	MSFC/STTE	19.12.67.	1640:21	Calibration	35.392s ib/35.508s ob	H-7092	H-7095	H-7086	H-7102	H-4092	H-4093	H-4094	H-4095	Successful
S-IB-11	SA-49	MSFC/STTE	25.1.68.	1707:45	15s	15.528s ib/15.644s ob	H-T6-B	H-7095	H-7086	H-7102	H-4092	H-4093	H-4067	H-4095	Successful bomb test
S-IB-11	SA-50	MSFC/STTE	6.2.68.	1640:02	15s	15.460s ib/15.576s ob	H-T6-B	H-7095	H-7086	H-7102	H-4092	H-4093	H-4067	H-4095	Successful bomb test
S-IB-11	SA-51	MSFC/STTE	14.2.68.	1630:01	15s	15.200s ib/15.320s ob	H-T6-B	H-7095	H-7086	H-7102	H-4092	H-4093	H-4067	H-4095	Successful bomb test
S-IB-11	SA-52	MSFC/STTE	21.2.68.	1640:00	15s	3.368s ib/3.484s ob	H-T6-B	H-7095	H-7086	H-7102	H-4092	H-4093	H-4067	H-4095	LOX seal leak caused explosion
S-IB-11	SA-53	MSFC/STTE	9.4.68.	1640:00	Calibration	35.304s ib/35.420s ob	H-7092	H-7095	H-7086	H-7102	H-4092	H-4093	H-4094	H-4091	Successful
S-IB-11	SA-54	MSFC/STTE	23.4.68.	1640:00	Full duration	142.332s ib/145.328s ob	H-7092	H-7095	H-7086	H-7102	H-4092	H-4093	H-4094	H-4091	Successful
S-IB-12	SA-55	MSFC/STTE	10.7.68.	1652	Calibration	35.4s	H-7100	H-7101	H-7098	H-7103	H-4073	H-4097	H-4098	H-4099	Successful
S-IB-12	SA-56	MSFC/STTE	25.7.68.	1640	Full duration	145.4s	H-7100	H-7101	H-7098	H-7103	H-4073	H-4097	H-4098	H-4099	Successful

S-IV Tests

Stage	Firing no.	Test stand	Date	Local Time	Duration Actual	Engines 201	202	203	204	205	206	Comments
S-IV-5	67300	SACTO/Alpha TS2B	5.8.63.		63.6s	P6418xx	P6418xx	P6418xx	P6418xx	P6418xx	P6418xx	Aborted due to fire indications
S-IV-5	67301	SACTO/Alpha TS2B	12.8.63.	1305	476.4s	P6418xx	P6418xx	P6418xx	P6418xx	P6418xx	P6418xx	Successful
S-IV-6	67275	SACTO/Alpha TS2B	22.11.63.		461s	P6418xx	P6418xx	P6418xx	P6418xx	P6418xx	P6418xx	Successful
S-IV-7	67336	SACTO/Alpha TS2B	29.4.64.		485s	P6418xx	P6418xx	P6418xx	P6418xx	P641849	P6418xx	Successful
S-IV-8		SACTO/Alpha TS2B	20.11.64.		475.8s	P641855	P641856	P641863	P641860	P641861	P641862	Successful
S-IV-9		SACTO/Alpha TS2B	6.8.64.		398.94s	P641857	P641850	P641851	P641852	P641836	P641854	Aborted due to low facility water pressure
S-IV-10		SACTO/Alpha TS2B	21.1.65.		479.50s	P641864	P641817	P641865	P641869	P641884	P641886	Successful

Saturn Missions

Mission ident	Alt name	First stage	Engines 101	102	103	104	105	106	107	108	Second stage	Engines 201	202	203	204	205	206	IU	SLA	Payload	Launch date	Launch time	Mission
SA-1		S-I-1	H-1016	H-1017	H-1019	H-1021	H-1011	H-1012	H-1013	H-1015											27.10.61.	1006:04	Sub-orbital test
SA-2		S-I-2	H-1032	H-1033	H-1034	H-1038	H-1028	H-1029	H-1035	H-1036											25.4.62.	0900:34	Project Highwater
SA-3		S-I-3	H-1045	H-1047	H-1048	H-1049	H-1040	H-1041	H-1042	H-1043											16.11.62.	1245:02	Project Highwater
SA-4		S-I-4	H-1054	H-1055	H-1056	H-1057	H-1051	H-1052	H-1053	H-1058											28.3.63.	1511:55	Engine out test
SA-5		S-I-5	H-5002	H-5003	H-5004	H-5005	H-2001	H-2002	H-2003	H-2004	S-IV-5	P6418xx	P6418xx	P6418xx	P6418xx	P6418xx	P6418xx	S-IU-5			29.1.64.	1125:01	Live 2nd stage. Orbital test
SA-6		S-I-6	H-5007	H-5008	H-5009	H-5010	H-2005	H-2006	H-2008	H-2007	S-IV-6	P6418xx	P6418xx	P6418xx	P6418xx	P6418xx	P6418xx	S-IU-6		CM BP-13	28.5.64.	1207:00	First BP Apollo
SA-7		S-I-7	H-5013	H-5027	H-5015	H-5016	H-2010	H-2019	H-2012	H-2021	S-IV-7	P6418xx	P6418xx	P6418xx	P6418xx	P641849	P6418xx	S-IU-7		CM BP-15	18.9.64.	1122:43	First programmable computer
SA-8		S-I-8	H-5019	H-5020	H-5021	H-5032	H-2016	H-2032	H-2018	H-2031	S-IV-8	P641855	P641856	P641863	P641860	P641861	P641862	S-IU-8		CM BP-26/Pegasus B	25.5.65.	0335:01	Pegasus B satellite
SA-9		S-I-9	H-5023	H-5012	H-5025	H-5026	H-2020	H-2022	H-2023	H-2024	S-IV-9	P641857	P641850	P641851	P641852	P641836	P641854	S-IU-9		CM BP-16/Pegasus A	16.2.65.	0937:03	Pegasus A satellite
SA-10		S-I-10	H-5028	H-5029	H-5030	H-5031	H-2034	H-2026	H-2027	H-2030	S-IV-10	P641864	P641817	P641865	P641869	P641884	P641886	S-IU-10		CM BP-9A/Pegasus C	30.7.65.	0800:00	Pegasus C satellite
SA-201		S-IB-1	H-7046	H-7047	H-7048	H-7049	H-4044	H-4045	H-4052	H-4047	S-IVB-201	J-2015						S-IU-201	SLA-3	CSM-009	26.2.66.	1112:01	Sub-orbital CM re-entry test
SA-202		S-IB-2	H-7051	H-7052	H-7050	H-7054	H-4048	H-4049	H-4050	H-4046	S-IVB-202	J-2016						S-IU-202	SLA-4	CSM-011	25.8.66.	1315:32	Sub-orbital CM re-entry test
SA-203		S-IB-3	H-7056	H-7058	H-7059	H-7060	H-4053	H-4054	H-4056	H-4057	S-IVB-203	J-2019						S-IU-203			5.7.66.	0953:17	Zero-g hydrogen experiment
SA-204	Apollo 5/LM-1	S-IB-4	H-7062	H-7063	H-7064	H-7065	H-4058	H-4062	H-4060	H-4061	S-IVB-204	J-2025						S-IU-204	SLA-7	LM-1	22.1.68.	1748:08	First lunar module flight
SA-205	Apollo 7	S-IB-5	H-7066	H-7067	H-7068	H-7069	H-4063	H-4064	H-4065	H-4066	S-IVB-205	J-2033						S-IU-205	SLA-5	CSM-101	11.10.68.	1102:45	First manned Apollo flight
SA-206	Skylab 2	S-IB-6	H-7071	H-7072	H-7073	H-7075	H-4068	H-4069	H-4070	H-4072	S-IVB-206	J-2046						S-IU-206	SLA-6A	CSM-116	25.5.73.	0900:00	First crew transfer to Skylab
SA-207	Skylab 3	S-IB-7	H-7085	H-7078	H-7076	H-7074	H-4078	H-4074	H-4075	H-4076	S-IVB-207	J-2056						S-IU-208	SLA-23	CSM-117	28.7.73.	0710:50	Second crew transfer to Skylab
SA-208	Skylab 4	S-IB-8	H-7082	H-7079	H-7081	H-7096	H-4077	H-4071	H-4079	H-4080	S-IVB-208	J-2062						S-IU-207	SLA-24	CSM-118	16.11.73.	0901:23	Third crew transfer to Skylab
SA-210	ASTP	S-IB-10	H-7091	H-7099	H-7093	H-7094	H-4087	H-4104	H-4089	H-4090	S-IVB-210	J-2087						S-IU-210	SLA-18	CSM-111	15.7.75.	1550:00	First US-USSR docking

S-IVB Tests

Stage	Firing no.	Test stand	Date	Local Time	Duration Actual	Engine 201	Comments
S-IVB-201		SACTO/Beta III	31.7.65.		0s	J-2015	Aborted due to instrumentation problem
S-IVB-201		SACTO/Beta III	8.8.65.		452s	J-2015	Successful
S-IVB-202		SACTO/Beta III	2.11.65.		0.41s	J-2016	Aborted due to instrumentation problem
S-IVB-202		SACTO/Beta III	9.11.65.		307s	J-2016	Aborted due to PU problem
S-IVB-202		SACTO/Beta III	1.12.65.		463.8s	J-2016	Successful
S-IVB-203		SACTO/Beta I	26.2.66.		284.9s	J-2019	Successful
S-IVB-204		SACTO/Beta III	18.3.66.		451.2s	J-2025	Successful
S-IVB-205		SACTO/Beta III	2.6.66.		437.5s	J-2033	Successful
S-IVB-206	614070	SACTO/Beta III	19.8.66.	1524:38	433.7s	J-2046	Successful but turbo-pump contamination
S-IVB-206	614072	SACTO/Beta III	14.9.66.		66.6s	J-2046	Turbo-pump calibration
S-IVB-207	614074	SACTO/Beta I	19.10.66.	1118:46	445.6s	J-2056	Successful
S-IVB-208	614076	SACTO/Beta I	12.1.67.	1212:16	426.6s	J-2062	Successful
S-IVB-209	614085	SACTO/Beta I	20.6.67.	1142:06	455.95s	J-2083	Successful

S-IVB MSFC Battleship

Firing no.	Date	Duration	Engine	Comments
S-IVB-001	2.8.65.	2.1s	J-2013	Scheduled 8 second test erroneously cut by a redline observer
S-IVB-002	10.8.65.	9.0s	J-2013	All test objectives met
S-IVB-003	18.8.65.	25.5s	J-2013	Gas generator and fuel turbopump replaced post-test due to GG deterioration and resultant overheating
S-IVB-004	8.9.65.	10.17s	J-2013	Planned duration was 80s. Erroneous cutoff by a redline observer
S-IVB-005	10.9.65.	80.0s	J-2013	As planned
S-IVB-006	15.9.65.	400.04s	J-2013	Normal operation
S-IVB-007	29.10.65.	250.0s	J-2027	S-IVB LOX tank pressurization system used for the first time. Gimballing not accomplished due to failure
S-IVB-008	16.11.65.	418.24s	J-2027	Gimballing and PU activation accomplished alternatively
S-IVB-009	23.11.65.	300.0s	J-2027	Gimbal test, PU system not activated
S-IVB-010	8.12.65.	388.90s	J-2027	Gimballing and PU activation accomplished simultaneously
S-IVB-011	17.12.65.	432.40s	J-2027	Evaluated performance of PU system, demonstrated gimballing performance
S-IVB-012	19.1.66.	7.3s		Redline cut off when the LOX pump temperature measurement device failed
S-IVB-013	24.1.66.	100.46s		Cut off due to overheated diffuser when the diffuser coolant water supply valve failed to open
S-IVB-014	26.1.66.	438.0s		All engine and stage parameters normal
S-IVB-015	4.2.66.	445.0s		All engine and stage parameters normal
S-IVB-016	21.2.66.	40.19s		Cut off by the gas generator over-temperature device
S-IVB-017	1.3.66.	412.44s		Cut off due to fuel depletion
S-IVB-018	9.3.66.	2.15s		Cut off by the gas generator over-temperature device
S-IVB-019	7.4.66.	200s		
S-IVB-020	13.4.66.	200s		
S-IVB-021	28.4.66.	200s		
S-IVB-022	?	?		
S-IVB-023	2.5.66.	151s		
S-IVB-024	4.5.66.	200s		
S-IVB-025	6.5.66.	323s		
S-IVB-026	16.5.66.	Ignition		
S-IVB-027	18.5.66.	11s	J-2048	Rough combustion with engine J-2048
S-IVB-028	10.6.66.	Ignition		
S-IVB-029	10.6.66.	Ignition		
S-IVB-030	26.7.66.	6.0 s	J-2048	No evidence of rough combustion
S-IVB-031	26.7.66.	200.5 s		No evidence of rough combustion
S-IVB-032	11.8.66.	31.0 s		Cut off by observer because of fire in the fuel turbopump area
S-IVB-033	24.8.66.	20.0 s		Cut off resulted when LOX pump inlet pressure exceeded redline limit
S-IVB-034	26.8.66.	279.0 s		Cut off by LOX depletion - hydraulic system operation successful, "hot gimballing"
				"Ignition Detected" signal not received. Test terminated by engine control package due to faulty ignition probe
S-IVB-035	9.9.66.	1.5 s		
S-IVB-036	9.9.66.	3.5 s		Test terminated by automatic vibration safety cut off device
S-IVB-037	5.10.66.	285.3 s		Cut off initiated by LOX Depletion System as planned
S-IVB-038	12.10.66.	234.0 s		Cut off by Fuel Depletion Cut off System. All test objectives were met
S-IVB-039	14.11.66.	288.4 s		Cut off was initiated by fuel level chart observer
S-IVB-040	13.12.66.	2.15 s		Cut off was initiated by automatic gas generator over temperature device because of erroneous indication due to faulty "drag in " instrumentation cable
S-IVB-041A	20.12.66.	153.01 s		Cut off was initiated by control after successful run for intended first burn duration
S-IVB-041B	20.12.66.	281.14 s		Fuel mass observer initiated depletion cut off
S-IVB-042	21.4.67.	235.8 s		After a simulated coast period of approx two hours this burner test was conducted. The LOX and LH2 tanks were repressurised, utilising the burner repressurisation systems. All parameters were normal
S-IVB-043	2.5.67.	150.0 s		All parameters were normal. This test demonstrated no performance variation due to PU valve
S-IVB-044	18.5.67.	4.7 s		Test terminated by an observer due to unusual appearance of the start transient
S-IVB-045	26.6.67.	199.0 s		Normal cut off on schedule
S-IVB-046	6.7.67.	436.39 s		All parameters were normal. All planned objectives were met
S-IVB-047	17.1.68.		J-108	J-2S firing
S-IVB-048	19.1.68.		J-108	J-2S firing
S-IVB-049	23.1.68.		J-108	J-2S firing
S-IVB-050	1.2.68.	74s	J-108	J-2S firing
S-IVB-051	8.2.68.	74+18s	J-108	J-2S restart firings
S-IVB-052	16.2.68.	200s	J-108	J-2S firing
S-IVB-053	1.3.68.		J-108	J-2S firing
S-IVB-054	7.3.68.		J-108	J-2S firing
S-IVB-055	14.3.68.		J-108	J-2S firing
S-IVB-056	8.5.68.	75s	J-2050	PU valve operation objectives not met
S-IVB-057	19.6.68.		J-2050	New ASI line assembly check
S-IVB-058	19.6.68.		J-2050	New ASI line assembly check
S-IVB-059	21.6.68.		J-2050	New ASI line assembly check
S-IVB-060	26.6.68.		J-2050	New ASI line assembly check
S-IVB-061	11.9.68.	100s		
S-IVB-062	19.9.68.	100s		
S-IVB-063	25.9.68.	aborted		Planned for 400s - terminated when fire observed in the main fuel valve area
S-IVB-064	8.10.68.	81.7s		
S-IVB-065	15.10.68.	354.7s		
S-IVB-066	22.10.68.	386.7s		
S-IVB-067	26.11.68.	385s		
S-IVB-068	4.12.68.	122.5s		Test in support of AS-503 launch
S-IVB-069	11.12.68.	435.4s		
S-IVB-070	14.12.68.	395.4s		

Saturn Transport Details

Primary cargo	Secondary cargo	Depart	Port	Arrive	Port	Transport
S-I-T1		14.3.61.	MSFC	17.3.61.	MSFC	Palaemon
S-I-T4.5		~12.62.	MSFC	~12.62	MAF	?
S-I-T4.5		~64.	MAF	~64.	MSFC	?
S-I-1	Dummy S-IV, dummy payload	5.8.61.	MSFC	5.8.61.	Wheeler Dam	Palaemon
S-I-1	Dummy S-IV, dummy payload	5.8.61.	Wheeler Dam	15.8.61.	Cape Canaveral	Compromise
S-I-2		17.2.62.	MSFC	17.2.62.	Wheeler Dam	Palaemon
S-I-2	Dummy S-IV, dummy S-V, dummy payload	17.2.62.	Wheeler Dam	27.2.62.	Cape Canaveral	Promise
S-I-3	Dummy S-IV, dummy S-IV	9.9.62.	MSFC	19.9.62.	Cape Canaveral	Promise
S-I-4	Dummy S-IV, dummy S-V, dummy payload, water ballast tank	20.1.63.	MSFC	2.2.63.	Cape Canaveral	Promise
S-I-5	S-IU-5, SA-5 payload	11.8.63.	MSFC	21.8.63.	Cape Canaveral	Promise
S-I-6	S-IU-6	7.2.64.	MSFC	18.2.64.	Cape Kennedy	Promise
S-I-7	S-IU-7	28.5.64.	MSFC	7.6.64.	Cape Kennedy	Promise
S-I-8		17.4.64.	MAF	25.4.64.	MSFC	Promise
S-I-8		24.6.64.	MSFC	29.6.64.	MAF	Promise
S-I-8		22.2.65.	MAF	28.2.65.	Cape Kennedy	
S-I-9	S-I fins, S-IU-9	19.10.64.	MSFC	30.10.64.	Cape Kennedy	Promise
S-I-10		24.7.64.	MAF	31.7.64.	MSFC	Promise
S-I-10		2.11.64.	MSFC	7.11.64.	MAF	Compromise
S-I-10		26.5.65.	MAF	31.5.65.	Cape Kennedy	Promise
S-I-D5		5.4.63.	MSFC	15.4.63.	Cape Canaveral	Promise
S-I-D5		1.7.63.	Cape Canaveral	14.7.63.	MSFC	Palaemon
S-I-D9			MSFC	22.7.64.	MAF	
S-IB-D/F		22.12.64.	MAF	4.1.65.	MSFC	Palaemon
S-IB-D/F			MSFC		MAF	
S-IB-D/F		23.9.66.	MSFC		MSFC	Promise
S-IB-D/F		28.6.69.	MSFC	28.6.69.	ASRC	Truck
S-IB-1		6.3.65.	MAF	14.3.65.	MSFC	Promise
S-IB-1		20.4.65.	MSFC	24.4.65.	MAF	Palaemon
S-IB-1	S-IU-200F/500F	9.8.65.	MAF	14.8.65.	Cape Kennedy	Promise
S-IB-2		12.6.65.	MAF	19.6.65.	MSFC	
S-IB-2		30.7.65.	MSFC	7.8.65.	MAF	Palaemon
S-IB-2		1.2.66.	MAF	7.2.66.	Cape Kennedy	Promise
S-IB-3		9.9.65.	MAF	16.9.65.	MSFC	Palaemon
S-IB-3		4.11.65.	MAF	9.11.65.	MAF	Barge
S-IB-3		7.4.66.	MAF	12.4.66.	Cape Kennedy	Promise
S-IB-4		7.12.65.	MAF	13.12.65.	MSFC	Palaemon
S-IB-4		29.1.66.	MSFC		MAF	Palaemon
S-IB-4		10.8.66.	MAF	15.8.66.	Cape Kennedy	Promise
S-IB-5			MAF	27.2.66.	MSFC	Palaemon
S-IB-5			MSFC		MAF	
S-IB-5		25.3.68.	MAF	28.3.68.	Cape Kennedy	Point Barrow
S-IB-6		19.5.66.	MAF	28.5.66.	MSFC	
S-IB-6		8.7.66.	MSFC	13.7.66.	MAF	
S-IB-6		13.12.66.	MAF	18.12.66.	Cape Kennedy	Palaemon
S-IB-6		3.4.67.	Cape Kennedy	10.4.67.	MAF	Promise
S-IB-6		17.8.72.	MAF	22.8.72.	Cape Kennedy	Orion
S-IB-7		4.8.66.	MAF	11.8.66.	MSFC	Poseidon
S-IB-7		20.9.66.	MSFC	25.9.66.	MAF	Palaemon
S-IB-7		24.3.73.	MAF	30.3.73.	Cape Kennedy	
S-IB-8		17.10.66.	MAF	25.10.66.	MSFC	
S-IB-8		8.12.66.	MAF	14.12.66.	MSFC	
S-IB-8		15.6.73.	MAF	20.6.73.	Cape Kennedy	
S-IB-9		19.1.67.	MAF	25.1.67.	MSFC	Palaemon
S-IB-9		15.3.67.	MSFC	20.3.67.	MAF	Palaemon
S-IB-9		14.8.73.	MAF	20.8.73.	Cape Kennedy	
S-IB-10		31.3.67.	MAF	7.4.67.	MSFC	Palaemon
S-IB-10		8.6.67.	MSFC	13.6.67.	MAF	Palaemon
S-IB-10	FWV H-1 engine	19.4.74.	MAF	24.4.74.	Cape Canaveral	
S-IB-11		20.10.67.	MAF	27.10.67.	MSFC	Palaemon
S-IB-11		11.5.68.	MSFC	16.5.68.	MAF	Palaemon
S-IB-11		~76.	MAF	~76.	MSFC	
S-IB-11		7.79.	MSFC	7.79.	I-65	
S-IB-12		23.4.68.	MAF	4.5.68.	MSFC	Palaemon
S-IB-12		7.8.68.	MSFC	12.8.68.	MAF	
S-IB-12		17.7.70.	MAF		MSFC	
S-IB-12		1.10.71.	MSFC	7.10.71.	MAF	
S-IB-12		8.5.74.	MAF		Cape Canaveral	
S-IB-13, S-IB-14		17.7.70.	MAF	28.7.70.	MSFC	
S-IB-13, S-IB-14		8.2.72.	MSFC	14.2.72.	MAF	
S-IV Battleship		12.61.	San Pedro	12.61.	Courtland Dock	Open deck barge
S-IV Battleship		12.61.	Courtland Dock	12.61.	SACTO	Truck
S-IV Battleship tank		21.5.63.	SACTO		Courtland Dock	Truck
S-IV Battleship tank			Courtland Dock		New Orleans	
S-IV Battleship tank			New Orleans	7.7.63.	MSFC	
S-IV-D5 (Hydrostatic/Dynamics vehicle)		26.10.62.	San Pedro		New Orleans	Smith Builder
S-IV-D5 (Hydrostatic/Dynamics vehicle)			New Orleans	16.11.62.	MSFC	Promise
S-IV-D9 (Hydrostatic/Dynamics vehicle)		28.6.69.	MSFC	28.6.69.	ASRC	Truck

Saturn Transport Details (cont)

Primary cargo	Secondary cargo	Depart	Port	Arrive	Port	Transport
S-IV Dynamics/Facilities vehicle						
		18.1.63.	San Pedro	1.2.63.	Cape Canaveral	
S-IV Dynamics/Facilities vehicle						
		1.8.63.	Cape Canaveral	2.8.63.	Los Angeles Airport	Pregnant Guppy
S-IV Dynamics/Facilities vehicle						
			Los Angeles Airport		Redstone Arsenal Airfield	Pregnant Guppy
S-IV ASV		1.2.63.	San Pedro		Courtland Dock	Open deck barge
S-IV ASV			Courtland Dock		SACTO	Truck
S-IV-5		15.4.63.	San Pedro		Mare Island Naval Shipyard	
S-IV-5			Mare Island Naval Shipyard		Courtland Dock	
S-IV-5			Courtland Dock	21.4.63.	SACTO	Truck
S-IV-5		20.9.63.	Mather Air Force Base	21.9.63.	Cape Canaveral	Pregnant Guppy
S-IV-6		27.9.63.	Santa Monica Airport	27.9.63.	Mather Air Force Base	Pregnant Guppy
S-IV-6		21.2.64.	Mather Air Force Base	22.2.64.	Cape Kennedy	Pregnant Guppy
S-IV-7		13.2.64.	Santa Monica Airport	13.2.64.	Mather Air Force Base	Pregnant Guppy
S-IV-7		10.6.64.	Mather Air Force Base	12.6.64.	Cape Kennedy	Pregnant Guppy
S-IV-8		7.8.64.	Santa Monica Airport	7.8.64.	Mather Air Force Base	Pregnant Guppy
S-IV-8		23.2.65.	Mather Air Force Base	26.2.65.	Cape Kennedy	Pregnant Guppy
S-IV-9		8.5.64.	Santa Monica Airport	8.5.64.	Mather Air Force Base	Pregnant Guppy
S-IV-9		21.10.64.	Mather Air Force Base	21.10.64.	Long Beach Airport	Pregnant Guppy
S-IV-9		22.10.64.	Long Beach Airport	22.10.64.	Cape Kennedy	Pregnant Guppy
S-IV-10			Santa Monica Airport	5.11.64.	Mather Air Force Base	Pregnant Guppy
S-IV-10			Mather Air Force Base	10.5.65.	Cape Kennedy	Pregnant Guppy
S-IVB Battleship		1.66.	Courtland Dock	8.1.66.	South Pittsburg	Barge
S-IVB Battleship		8.1.66.	South Pittsburg	8.1.66.	AEDC	Truck
S-IVB-D		9.12.64.	Seal Beach Naval Weapons Station	21.12.64.	New Orleans	Aloha State
S-IVB-D		21.12.64.	New Orleans	22.12.64.	MAF	Promise
S-IVB-D		22.12.64.	MAF	4.1.65.	MSFC	Promise
S-IVB-D		20.3.66.	MSFC	20.3.66.	Los Alamitos Naval Air Station	Super Guppy
S-IVB-D		24.3.66.	Los Alamitos Naval Air Station	24.3.66.	MSFC	Super Guppy
S-IVB-D		28.6.69.	MSFC	28.6.69.	ASRC	Truck
S-IVB-F		12.2.65.	Seal Beach Naval Weapons Station	17.2.65.	Courtland Dock	Orion
S-IVB-F		17.2.65.	Courtland Dock	17.2.65.	SACTO	Truck
S-IVB-F		10.6.65.	SACTO	10.6.65.	Courtland Dock	Truck
S-IVB-F		10.6.65.	Courtland Dock	13.6.65.	Seal Beach Naval Weapons Station	Orion
S-IVB-F	S-II simulator	13.6.65.	Seal Beach Naval Weapons Station	26.6.65.	MAF	Point Barrow
S-IVB-F		26.6.65.	MAF	30.6.65.	Cape Kennedy	Point Barrow
S-IVB-F		69.	Cape Kennedy	69.	MSFC	?
S-IVB-F		2.1.70.	MSFC	2.1.70.	Los Alamitos Naval Air Station	Super Guppy
S-IVB-F	S-IVB-512	4.12.70.	Seal Beach Naval Weapons Station	18.12.70.	MAF	Point Barrow
S-IVB-F		31.12.70.	MAF	5.1.71.	MSC	Orion
S-IVB-F		23.5.71.	MSC	4.6.71.	MSFC	Orion
S-IVB-F		6.74.	MSFC	6.74.	Cape Canaveral	
S-IVB-201		30.4.65.	Seal Beach Naval Weapons Station	5.5.65.	Courtland Dock	Orion
S-IVB-201		6.5.65.	Courtland Dock	6.5.65.	SACTO	Truck
S-IVB-201		3.9.65.	SACTO	3.9.65.	Courtland Dock	Truck
S-IVB-201		3.9.65.	Courtland Dock		Mare Island Naval Shipyard	Orion
S-IVB-201			Mare Island Naval Shipyard	19.9.65.	Cape Kennedy	Steel Executive
S-IVB-202		28.8.65.	Seal Beach Naval Weapons Station	1.9.65.	Courtland Dock	Orion
S-IVB-202		1.9.65.	Courtland Dock	1.9.65.	SACTO	Truck
S-IVB-202		15.1.66.	SACTO	15.1.66.	Courtland Dock	Truck
S-IVB-202		15.1.66.	Courtland Dock		Mare Island Naval Shipyard	Orion
S-IVB-202	S-IVB-202 aft inter-stage, S-IV mock-up		Mare Island Naval Shipyard	31.1.66.	Cape Kennedy	Point Barrow
S-IVB-203		29.10.65.	Seal Beach Naval Weapons Station	1.11.65.	Courtland Dock	Orion
S-IVB-203		1.11.65.	Courtland Dock	1.11.65.	SACTO	Truck
S-IVB-203		4.4.66.	Mather Air Force Base	6.4.66.	Cape Kennedy	Super Guppy
S-IVB-204		10.1.66.	Seal Beach Naval Weapons Station	14.1.66.	Courtland Dock	Orion
S-IVB-204		14.1.66.	Courtland Dock	14.1.66.	SACTO	Truck
S-IVB-204		6.8.66.	Mather Air Force Base	6.8.66.	Cape Kennedy	Super Guppy
S-IVB-205		9.4.66.	Seal Beach Naval Weapons Station	13.4.66.	Courtland Dock	Orion
S-IVB-205		13.4.66.	Courtland Dock	13.4.66.	SACTO	Truck
S-IVB-205		6.4.68.	Mather Air Force Base	8.4.68.	Cape Kennedy	Super Guppy
S-IVB-206		30.6.66.	Los Alamitos Naval Air Station	1.7.66.	Mather Air Force Base	Super Guppy
S-IVB-206		13.12.66.	Mather Air Force Base	14.12.66.	Cape Kennedy	Super Guppy
S-IVB-206		13.4.67.	Cape Kennedy	14.4.67.	Mather Air Force Base	Super Guppy
S-IVB-206		3.8.70.	Mather Air Force Base		Los Alamitos Naval Air Station	Super Guppy
S-IVB-206		23.6.71.	Los Alamitos Naval Air Station	24.6.71.	Cape Kennedy	Super Guppy
S-IVB-207		30.8.66.	Los Alamitos Naval Air Station	31.8.66.	Mather Air Force Base	Super Guppy
S-IVB-207		1.5.70.	Mather Air Force Base	1.5.70.	Los Alamitos Naval Air Station	Super Guppy
S-IVB-207		25.8.71.	Los Alamitos Naval Air Station	26.8.71.	Cape Kennedy	Super Guppy
S-IVB-208		2.12.66.	Los Alamitos Naval Air Station	2.12.66.	Mather Air Force Base	Super Guppy
S-IVB-208		13.10.70.	Mather Air Force Base	13.10.70.	Los Alamitos Naval Air Station	Super Guppy
S-IVB-208		3.11.71.	Los Alamitos Naval Air Station	5.11.71.	Cape Kennedy	Super Guppy
S-IVB-209		9.3.67.	Los Alamitos Naval Air Station	9.3.67.	Mather Air Force Base	Super Guppy
S-IVB-209		22.7.70.	Mather Air Force Base	22.7.70.	Los Alamitos Naval Air Station	Super Guppy
S-IVB-209		11.1.72.	Los Alamitos Naval Air Station	12.1.72.	Cape Kennedy	Super Guppy
S-IVB-210		7.11.72.	Los Alamitos Naval Air Station	8.11.72.	Cape Kennedy	Super Guppy
S-IVB-211		17.10.68.	Los Alamitos Naval Air Station	18.10.68.	Mather Air Force Base	Super Guppy
S-IVB-211		15.9.70.	Mather Air Force Base	15.9.70.	Los Alamitos Naval Air Station	Super Guppy
S-IVB-211		26.6.72.	Los Alamitos Naval Air Station	27.6.72.	Cape Kennedy	Super Guppy
S-IVB-211			Cape Kennedy		Redstone Arsenal Airfield	Super Guppy?
S-IVB-211			MSFC		ASRC	Truck
S-IVB-212 (Skylab)	Payload shroud	8.9.72.	Seal Beach Naval Weapons Station	22.9.72.	Cape Kennedy	Point Barrow

References

History of the George C Marshall Space Flight Center from January 1 through June 30 1961 volume 1. NASA MHM-3. November 1961.
History of the George C Marshall Space Flight Center from January 1 through June 30 1961 volume 2. NASA MHM-3. November 1961.
History of the George C Marshall Space Flight Center from July 1 through December 31 1961 volume 1. NASA MHM-4. March 1962.
History of the George C Marshall Space Flight Center from July 1 through December 31 1961 volume 2. NASA MHM-4. March 1962.
History of the George C Marshall Space Flight Center from January 1 through June 30 1962 volume 1. NASA MHM-5. September 1962
History of the George C Marshall Space Flight Center from January 1 through June 30 1962 volume 2. NASA MHM-5. September 1962.
History of the George C Marshall Space Flight Center from July 1 through December 31 1962 volume 1. NASA MHM-6. May 1963.
History of the George C Marshall Space Flight Center from July 1 through December 31 1962 volume 2. NASA MHM-6. May 1963.
History of the George C Marshall Space Flight Center from January 1 through June 30 1963 volume 1. NASA MHM-7. November 1963.
History of the George C Marshall Space Flight Center from January 1 through June 30 1963 volume 2. NASA MHM-7. November 1963.
History of the George C Marshall Space Flight Center from July 1 through December 31 1963 volume 1. NASA MHM-8. July 1964.
History of the George C Marshall Space Flight Center from July 1 through December 31 1963 volume 2. NASA MHM-8. July 1964.
History of the George C Marshall Space Flight Center from January 1 through June 30 1964 volume 1. NASA MHM-9. May 1965.
History of the George C Marshall Space Flight Center from January 1 through June 30 1964 volume 2. NASA MHM-9. May 1965.
History of the George C Marshall Space Flight Center from July 1 through December 31 1964 volume 1. NASA MHM-10.
History of the George C Marshall Space Flight Center from July 1 through December 31 1964 volume 2. NASA MHM-10.
History of the George C Marshall Space Flight Center from January 1 through December 31 1965 volume 1. NASA MHM-11. April 1968.
History of the George C Marshall Space Flight Center from January 1 through December 31 1965 volume 2. NASA MHM-11. April 1968.
A chronology of the George C Marshall Space Flight Center from January 1 through December 31 1966. NASA MHR-6. November 1969.
A chronology of the George C Marshall Space Flight Center from January 1 through December 31 1967. NASA MHR-7. April 1970.
A chronology of the George C Marshall Space Flight Center from January 1 through December 31 1968. NASA MHR-8. February 1971.
A chronology of the George C Marshall Space Flight Center from January 1 through December 31 1969. NASA MHR-9 (draft never issued). June 1972.
An illustrated chronology of the NASA Marshall Center and MSFC programs 1960-1973. NASA MHR-10. May 1974.
Saturn illustrated chronology. April 1957 to April 1968. NASA MHR-5. 20 January 1971.
Saturn I/IB progress report. MPR-SAT I/IB-64-2&3. 16 March 1964 – 30 September 1964.
Dynamic test results of SAD-8, 9. NASA TM X-53188. 15 January 1965.
Test results part 1 of the firing test report Saturn vehicle SA-1. MTP-LOD-61-36.1. 8 November 1961.
Saturn SA-1 flight evaluation. MPR-SAT-WF-61-8. 14 December 1961.
Saturn SA-2 flight evaluation. MPR-SAT-WF-62-5. 5 June 1962.
Results of the third Saturn I launch vehicle test flight. MPR-SAT-64-13. 26 February 1964.
Results of the third Saturn I launch vehicle test flight. MRP-SAT-63-3. No date.
Results of the sixth Saturn I launch vehicle test flight. MPR-SAT-FE-64-18. 1 October 1964.
Evaluation of flight test propulsion systems and associated systems Saturn vehicle SA-6. IN-P&VE-P-64-19. No date.
Results of the seventh Saturn I launch vehicle test flight. MPR-SAT-FE-64-19. 30 December 1964.
Results of the ninth Saturn I launch vehicle test flight, SA-8. MPR-SAT-FE-66-10. 13 June 1966.
Results of the eighth Saturn I launch vehicle test flight, SA-9. MPR-SAT-FE-66-4. 28 February 1966.
Evaluation of flight test propulsion systems and associated systems Saturn S-I-9 stage. IN-P&VE-P-65-5. 22 April 1965.
Results of the tenth Saturn I launch vehicle test flight. MPR-SAT-FE-66-11. 14 July 1966.
Results of the first Saturn IB launch vehicle test flight. MPR-SAT-FE-66-8. 6 May 1966.
Saturn S-IB stage assembly and test report S-IB-1. RB-B1-EIR-5.1. 9 August 1965.
Results of the second Saturn IB launch vehicle test flight. AS203. MPR-SAT-FE-66-12. 22 September 1966.
Results of the fourth Saturn IB launch vehicle test flight. MPR-SAT-FE-68-2. 5 April 1968.
Results of the fifth Saturn IB launch vehicle test flight. MPR-SAT-FE-68-4. 25 January 1969.
Uprated Saturn I – S-IB stage. Supplement assembly and test report. RB-B5-EIR-5.1. 25 March 1968.
Saturn IB launch vehicle flight evaluation report SA 206. MPR-SAT-FE-73-3. 23 July 1973.
Skylab 2 post launch report. RCS 76-0000-00048. 21 June 1973.
Saturn IB launch vehicle flight evaluation report SA 207. MPR-SAT-FE-73-5. 8 October 1973.
Saturn IB launch vehicle flight evaluation report SA 208. MPR-SAT-FE-74-1. 31 January 1974.
Saturn S-IB stage final flight report S-IB-8. SDES-74-425. 25 January 1974.
Preliminary report – S-IV all systems stage incident. Dr K Debus. 19 February 1964.
First flight vehicle, S-IV-5. Douglas report.
Second flight vehicle, S-IV-6. Douglas report.
Third flight vehicle, S-IV-7. Douglas report.
Fourth flight vehicle, S-IV-9. Douglas report.

S-IV-10 stage acceptance firing report. SM-46970. April 1965.

Saturn S-IVB-204 stage flight evaluation report. SM 46989. April 1968.

Saturn S-IVB-205 stage flight evaluation report. SM 46990. December 1968.

Saturn S-IVB-206 stage acceptance firing report. SM 47472. October 1966.

Saturn S-IVB-207 stage acceptance firing report. SM 47473. December 1966.

Saturn S-IVB-208 stage acceptance firing report. SM 47474. 27 March 1967.

Saturn S-IVB-209 stage acceptance firing report. DAC 47475. August 1967.

Narrative end item report Saturn S-IVB-208. DAC-56499. June 1968.

Narrative end item report Saturn S-IVB-209. DAC-56502. August 1968.

Narrative end item report Saturn S-IVB-210. DAC-56503. April 1968.

Narrative end item report Saturn S-IVB-212. DAC-56581. September 1968.

Saturn S-1 stage final static test report, stage S-I-8. CCSD. No date.

Saturn S-1 stage final static test report, stage S-I-9. CCSD. No date.

Saturn S-1 stage final static test report, stage S-I-10. CCSD. No date.

Saturn S-1B stage final static test report, stage S-IB-1. CCSD. No date.

Saturn S-1B stage final static test report, stage S-IB-2. CCSD. No date.

Saturn S-1B stage final static test report, stage S-IB-4. CCSD. No date.

Saturn S-1B stage final static test report, stage S-IB-6. CCSD. No date.

Saturn S-1B stage final static test report, stage S-IB-8. CCSD. No date.

Saturn S-1B stage final static test report, stage S-IB-9. CCSD. No date.

Saturn S-1B stage final static test report, stage S-IB-10. CCSD. No date.

Saturn S-1B stage final static test report, stage S-IB-11. CCSD. No date.

Failure analysis engine H-4095. 27 May 1968.

H-1 engine LOX dome failure. C E Cataldo, MSFC. No date.

H-1 configuration identification & status report. R-7392. 3 January 1975.

H-1 engine status. MSFC list. 22 August 1974.

Untitled datasheets – listings of all firings at MSFC. December 29 1970.

Saturn S-IVB quarterly technical progress report. Douglas DAC-56533. March 1967.

Saturn/Apollo and MOL congressional record presentation. Douglas. 11 February 1966.

Sacramento Test Center resources handbook. Douglas SM 37538 R1. December 1966.

Transportation of Douglas Saturn S-IVB stages. Douglas 3688. November 1965.

S-IVB/IB and S-IVB/V common battleship report. Douglas SM-47012. 21 February 1966.

Saturn J-2 configuration identification & status report. Rocketdyne R-5788. 16 July 1975.

J-2 program monthly progress report for period ending 28 February 1962. R-2599-18. 7 March 1962.

J-2 program monthly progress report for period ending 30 June 1963. R-2599-34. 8 July 1963.

J-2 Propulsion Production System Data. Rocketdyne. No refs.

J-2 engine significant configuration change points. Internal letter, NAR. 9 March 1971.

MSFC test laboratory, monthly progress report, February 1, 1967 through February 28, 1967.

Contractor photographic coverage still and motion picture. Monthly reports from August 1967 to January 1971. MSFC.

Liquid Propellant Rocket Propulsion Systems, Rocketdyne brochure. 1999.

Marshall Space Flight Center. Historic Aerospace Site. AIAA. 2002.

Air Force Research Lab. Historic Aerospace site. AIAA. 2000.

The Rocketdyne Santa Susana Field Lab. Historic Aerospace site. AIAA. 2001.

NavSource Online: Service Ship Photo Archive. Web site.

All about Guppys. Daren Savage. Web site.

Cloudster. S-IVB photographs. Phil Broad. Web site.

MSFC Marshall Star. CD ROM of all past issues.

H-1 Rocket Engine, models H-1C and H-1D, Technical Manual Pocket Data Supplement, Rocketdyne R-3620-1A, dated 22 July 1971.

Saturn S-IV handbook, Douglas, first issue, October 1962.

Saturn Data Summary Handbook, Douglas, July 1964.

Mike Wright, MSFC historian, notes to file, 16 October 1996.

Notes to Brown. H-1 engine project. 9 December 1966.

Explosion at S-IV battleship stand. Memo E Rees to Col James, 7 July 1965.

Various memos, telexes, letters regarding transportation of S-IVB stages.

Various telexes regarding testing of S-IV stages at SACTO.

MSFC management council meeting minutes. 24 May 1962.

MSFC program status for November 27, 1962.

Basic information data, Michoud Ordnance Plant, New Orleans. Circa 1962.

Various S-IV progress memos from the SACTO resident.

Saturn technical review meeting. Douglas. 13 February 1963.

Saturn monthly technical progress and quarterly report. Douglas SM-46636. December 1963.

S-IV program briefing. Douglas. 10 April 1963.

Information about Pratt & Whitney Aircraft. P&W press release. 1965.

Spacetown USA. Premier Turbines. 2006.

History of liquid rocket engine development in the United States 1955-1980. AAS History Series, volume 13.

Saturn vehicle cryogenic programs. NASA TMX 57239. August 1965.

Development of LOX-hydrogen engines for the Saturn Apollo launch vehicles. A J Burks, MSFC. 10 June 1968.

Astronautics. A publication of the American rocket society. February 1962.

Recent NASA experience with hydrogen engines. MSFC. AIAA paper 64-270.

Development of LOX/RP-1 engines for Saturn/Apollo launch vehicles. L Bostwick. MSFC. AIAA paper 68-569.

Index

Photo credits:

Allen, 11, 12, 29 (upper)

Brincka, 18 (lower), 19, 20 (left), 21, 22 (left), 23 (left), 25 (upper), 60, 61 (upper), 126 (upper), 139, 140 (upper), 142, 143 (lower left), 144, 145 (lower), 152, 155, 161, 162, 163, 164, 165, 167

Broad, 61 (lower), 168 (upper), 169, 172, 173, 174, 175, 176 (upper left), 177 (right), 178, 179, 181, 182 (left)

Lawrie, 15, 20 (right middle, right lower), 22 (right), 23 (right upper), 24, 26, 59 (left), 93 (right), 111 (right), 117 (right), 118, 126, 128 (lower), 140 (lower), 151 (lower), 154, 159 (upper right and lower), 168 (lower left), 170, 176 (upper right and lower), 177 (left), 180 (upper right and lower)

MIX, 15 (upper), 30 (upper), 31 (lower right), 59 (lower), 64 (lower), 66 (lower), 68, 69 (right and lower), 70 (upper and left), 73, 74, 75 (left), 78, 79, 84 (upper), 85 (middle), 88, 90, 91, 92 (left), 94, 95 (upper), 98, 101, 105 (upper), 107, 113, 115, 116, 117 (lower left), 121 (upper), 122 (lower), 123, 124 (upper left and middle), 125 (left), 126 (lower), 145 (upper), 146 (lower), 148, 155, 159 (upper left), 182 (right)

NARA, 56, 58 (upper), 62, 63, 83, 84 (lower), 109, 117 (upper left), 121 (lower), 122 (upper), 124 (upper right and lower), 125 (right), 150, 168 (lower right), 180 (upper left)

NASA, 18 (upper), 64 (upper), 66 (upper), 67, 69 (upper left), 71, 75 (upper right), 77, 85 (upper and lower), 86, 89, 93 (left), 95 (lower), 128 (upper), 143 (lower right), 154, 155

Premier Turbines, 13, 14, 15 (middle)

Robinson, 127, 143 (upper), 146 (upper), 151 (upper)

Rocketdyne, 29 (lower)
Smyth, 110

Teague, 70 (lower right), 75 (lower), 105 (lower), 111 (left)
UAH, 58 (lower), 59 (upper right)

von Puttkamer, 92 (right)

Wheelock, 15 (lower), 16, 20 (upper), 23 (right lower), 25 (lower), 30 (lower), 31 (upper and lower left), 32, 47, 49, 50, 51, 61, 80, 97, 100, 102, 103, 104, 108, 112, 114, 171